©2013 by Norman Ahlhelm

Introduction

This book is an introduction to the techniques of electronic repair and, to a lesser extent, fabrication. Electronic repair in this text is directed only at soldering and board repair, not troubleshooting. And, as with all texts dealing with technology, the book is not inclusive of every method available to the technician. No prior knowledge of the operation of electronic components or repair expertise is needed. Basic soldering tools are used as much as possible to perform the work. Unfortunately, some specialized tools will have to be used for board repairs and the installation of surface mount components.

This edition has minor changes to include: metric wire gages, standards are now listed in metric also, improved some images, and revision of a few confusing questions.

The standard used throughout the text for all soldering is IPC (Association Connecting Electronics Industries, formerly the Institute of Interconnecting and Packaging Electronic Circuits) J-STD-001D dated February 2005. It is the accepted model used by the American National Standards Institute as well as the Department of Defense. Repair standards should meet IPC-7711/7721.

Safety

Listed here are some of the safety precautions that should be observed at all times. Additionally, follow the procedures established by Federal and State regulations and your employer.

1) Know your tools and supplies.

2) Inspect your tools to make sure they are in perfect working order.

3) Wear eye or face protection.

4) Work in a well-ventilated area and/or have fume extraction devices operating if available.

5) Wash your hands after working with solder, particularly those containing lead.

6) Avoid wearing loose jewelry or clothing – ties are out!

7) Wear long sleeves and pants – this prevents burns if solder should splatter.

8) No open toed shoes – same reason as six. You may also be working with heavy objects.

9) Do not touch the irons tip even if you think it is cold.

10) Keep your work area clean and organized.

11) Make sure the iron never touches the power cord.

12) Ergonomics must be considered. Make sure the chair is in the correct position, the work table is at a proper height, and lighting is sufficient.

13) Solder on fire resistant surfaces.

14) Do not overload your electrical outlets.

15) Shut your equipment down when not in use.

Table of Contents

Chapter 1, The Basics ....

    Conductors 6

    Wire Gage 7

    Insulation 10

    Skin Affect and Proximity Affect 12

    Other Conductors

        Waveguides 13

        Coaxial Cable 14

Chapter 2, Standards: Wires and Splices

    Wire Stripping 16

    How to Test for Proper Wire Stripping 19

    Wire Connections

        Appliance Splice 20

        Western Union Splice......................... 21

        Pig Tail 23

        Tap Splice 24

        Mesh Splice 25

        Lap Splice 26

    Crimping and Splicing Tools 26

    Solder Splicing Sleeves........................... 28

    Coaxial Cable Connectors 28

    Molex Connectors 32

    RJ45 Plugs and Jacks 34

Chapter 3, Irons, Solders, and Fluxes

    What is soldering? 41

    The Tools

        The Soldering Iron 41

        Solder 43

        Flux 46

    Classifying Types of Flux 50

    The How and Why of Soldering 51

    Mass, Time, and Pressure

        Mass – Work Piece Size 52

        Time 53

        Pressure 53

    Some Basic Tools 54

Chapter 4, Wire Connections and Connectors

    Tinning 57

    The Care and Feeding of the Soldering Iron Tip 58

    Preparing to Solder 59

    Tinning Wires 60

    Hooks and Eyelets 62

    The Turret 64

The Bifurcated Connector 66
Cups 67
Connectors with Gold Plating   70
Soldering Wires Splices
    The Western union Splice 70
    Sealing the Western Union Splice 71
    Soldering the Tap Splice 73
    Sealing the Tap Splice   73
Different Solder – Different Appearance 74

Chapter 5, Printed Circuit Boards
In the Beginning 75
Boards types 76
Making the Connection 78
Single-Side Boards 78
Double-Sided Boards   79
Multi-Layer Boards 80
The Artwork 81
Making a Circuit
    Subtractive Methods 82
        Photo Developing 82
        Silk Screening 85
        Milling 86
    Additive Methods 87
    Tinning 87
    Drilling of Through-Holes ...._............................... 89

Chapter 6, Installation and Removal of
    Through-Hole Components
What Does the Board Tell You? 91
Gravity, Surface Tension, and Capillary Action
    Gravity 92
    Surface Tension ...._.................. 93
    Capillary Action ...._.................. 93
Before Component Installation
    Lead Bending 94
    Lead forming 95
    Lead Forming Tools............... 96
Mounting an Unsupported
      Through-Hole Component 96
Mistakes in Through-Hole Soldering _................................. 98
Plated Through-Holes   ...._.............................. 99
Stress reliefs 101
Dual-In-Line Packages   (DIPS) 103
Heat Sinks 103
Through-Hole Component Removal
    Conformal Coatings.................._...................... 103

Removal Methods 105
Solder Wick 108
Air Impulse 109
Constant Vacuum 111

Chapter 7, Removal and Installation of
    Surface Mount Components
Surface Mount Component Types 113
Installation and removal of Chip Resistors, MELFs,
Chip Capacitors and SOTs
Chip Resistors 117
MELFs and Chip Capacitors 120
SOTs 121
Installation of QFPs, PLCCs, and SOICs
QFPs 122
SOICs and PLCCs 123
Removal of QFPs, PLCCs, and SOICs 127
Mixed Technology Boards 128
Ball Grid Array (BGA) 129
Preparation 129
Re-Balling Using Solder Paste 130
Re-Balling Using Solder Spheres 131

Chapter 8, Track, Pad, and Board Repairs
Damaged or Missing Tracks
Lifted Pads and Tracks 134
Broken or Missing Tracks
Using a Track Frame without Adhesive 134
Using a Track Frame with Adhesive 140
Busbars 142
Staples 143
Jumper Wires 144
Edge Connectors 147
Replacing Large Sections of Damaged PCBs
Small Repairs (1/2" or Less): Repair Does
Not Extend Through the Board............................ 150
Damage Extends Through the Board 151
Replacement of Large Sections of a PCB 152

Appendix A, Electrostatic Discharge (ESD)............................ 153

Chapter One

The Basics

A wire must be *electrically transparent* to the circuit in which it functions. By electrically transparent we mean that it must not affect the current or voltage of the circuit. So, how can a wire affect current or voltage in a poor design? How will a wire react to different frequencies? How does a wire function in a proper design? Finally, if you have to make a repair to a wire, how do you make it as mechanically strong as the original and make it electrically sound?

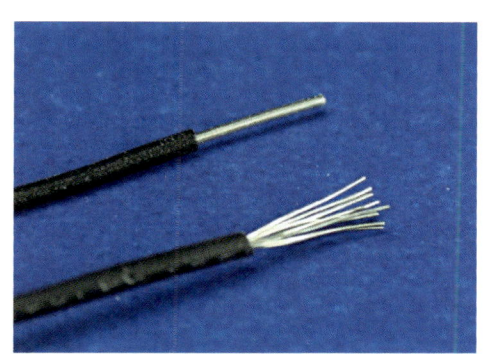

Figure 1-1
Solid and Stranded wires.

	Conductance Compared to
Material	**Silver**
Silver	100
Copper	98
Gold	78
Aluminum	61
Iron	16
Tin	9
Carbon	0.05

Table 1-1

Conductors

The most commonly used conductor is the wire. It comes in two varieties, **Stranded** and **Solid**. The material from which they are usually made is copper (because of its very low resistance and reasonable cost). **Figure 1-1** shows both a solid conductor and a stranded conductor. Each type of wire has certain benefits and shortcomings. The solid wire is easier to manufacture, it is only a single strand. Its disadvantage is that solid wire is not intended to be used in areas where it must be bent regularly. This wire may be bent only a few times before it breaks. Stranded wire, on the other hand, is more costly to manufacture: a certain number of strands of a given size must be wound together to achieve the desired diameter. It has the advantage of being more resilient than a solid wire (of similar size) when wires must be bent. It can be flexed more often without damage.

Copper, as mentioned earlier, is an excellent conductor. The resistance of copper per foot is extremely low, less than 0.077 ohms for a wire having a diameter of 0.012". This makes it largely transparent electrically since it does not affect current much or create voltage drops. As the size of the wire increases, this resistance value continues to drop. **Table 1-1** shows the relative conductance of copper as compared to the "ideal" material, silver. Silver might be the best conductor available, but it is too costly for most consumer electronics. Gold is also listed. You can see that it has 78% of the conductance of silver. Aluminum is rarely, if ever, used in circuits. Because of its considerable resistance, heat tends to build up along its length. Aluminum wire was used in residential electrical systems during the 1970's and was responsible for several home fires because of its high resistance.

Wire Gauge

The diameter of a conductor has to be sized to the application, some circuits require more current and voltage than others. In the United States, the American Wire Gauge (AWG) standard is used as a measure of conductor diameter. **Table 1-2** lists some of the common sizes used. Most manufacturers of electronics use the even numbered gauges so many of the odd number values have been omitted.

AWG	Metric Wire[3] (mm^2)	Diameter in mils	Area in cmils[1]	Ohms Per 1000 Ft	Maximum Current*	Fuse Ratings[2]
40		3.145	9.891	1049	13.7mA	1.77A
	0.09	3.543	12.555		17.9mA	
38		3.965	15.721	659.6	22.8mA	2.5A
	0.112	4.409	19.44		27.77mA	
36		5	25	428.2	34mA	3.62A
	0.14	5.512	30.379		43.4mA	
34		6.305	39.753	260.9	56mA	5.12A
	0.18	7.086	50.218		71mA	
32		7.95	63.203	164.1	91mA	7.19A
	0.25	9.842	96.872		138mA	
30		10.025	100.5	103.2	142mA	10.2A
	0.31	11.8	139.236		198mA	
28		12.641	159.795	64.9	226mA	14.4A
	0.355	13.976	195.333		279mA	
26		15.94	254.084	40.81	361mA	20.5A
	0.5	19.685	387.488		553mA	
24		20.1	404.01	25.67	577mA	29.2A

AWG	Metric Wire[3] (mm²)	Diameter in mils	Area in cmils[1]	Ohms Per 1000 Ft	Maximum Current*	Fuse Rating s[2]
	0.6	23.622	557.98		797mA	
22		25.347	642.47	16.14	920mA	41.2A
	0.7	31,417	987	11.03	1.4A	
20		31.961	1021.506	10.15	1.5A	58.4A
	0.75	38.471	1480	7.35	2A	
18		40.303	1524.735	6.385	2.3A	82.9A
	1 (7/.432)	42.43	1974	5.52	2.8A	
16		50.82	2582.672	4.016	3.7A	117A
	1.5 (7/.533)	54.41	2960	3.69	4.2A	
14		64.084	4106.759	2.525	5.9A	166A
	2.5 (7/.686)	70.24	4934	2.26	7A	
12		80.808	6529.933	1.588	9.3A	235A
	4 (7/.864)	88.848	7777.04	1.41	11.1A	
10		101.89	10381.57	0.9989	15A	333A
	6 (7/1.07)	104.33	10884.541		15.5A	
9		114.43	13094.23	0.7921	19A	
	9	118.108	13949.571		19.92A	
8		129.49	16767.66	0.6282	24A	
	10 (7/.978)	131.89	17394.341		24.8A	
7		144.28	20816.72	0.4982	30A	
6		162.02	26250.48	0.3951	37A	
5		181.94	33102.16	0.3133	47A	
4		204.31	41742.58	0.2485	60A	
3		229.42	52633.54	0.197	75A	
2		257.63	66373.22	0.1563	94A	
1		289.3	83694.49	0.1239	119A	
0		324.86	105534	0.0983	150A	

*700 cmils per amp rule operating in free air

(1) Metric cmil values are approximate

(2) These are general values for copper in free air. These values are approximations only! The values of ohms per foot, maximum current, and fuse rating are for copper wire only!

(3) Parenthesized values in this column are the metric equivalent for stranded wire. 7/.107 indicates 7 strands of 0.107mm wire which will be equivalent in circular area to a 1mm solid wire

---

Table 1-3

Figure 1-2
Various wire gauges achieved through the extrusion process.

Notice that the largest AWG, 40, has the smallest diameter (expressed in mils). A mil is one-thousandth of an inch so a 40 AWG wire is only 0.003145 inches across or 1/318th of an inch. As the gauge numbers become smaller, the diameter of the wire increases. A commonly used wire is 22 AWG, yet it is eight (8) times larger than the 40 AWG conductor. A wire of 0 AWG is about 1/3rd of an inch across. The reason the smaller gauges having a larger diameter stems from the manufacturing methods used. To create a wire of a given diameter, a copper rod is drawn or pulled through a die. The smaller the wire has to be, the more times the wire has to be drawn through progressively smaller dies. The AWG number was based on the number of times this extrusion process had to be repeated to achieve the desired thickness.

Solid wire size expressed in the metric system is easier to understand. Wire diameter is measured in millimeters (mm) of diameter (there are 25.4mm to the inch). It is apparent that a 2mm wire must have twice the width of a 1mm wire. A metric wire may also be expresses by cross-sectional area – just square the diameter in millimeters. A 0.7mm wire has a cross-sectional are of $0.49mm^2$. Metric wires can be identified another way – by their diameter or area as mentioned above or as a **Metric Gauge**. Metric Gauges are simply the diameter in millimeters multiplied by 10. More often than not, the metric wires are simply identified by their diameter in millimeters.

Stranded wire in the metric system is simply expressed as the number of strands and their diameter. A 7/0.017 is a stranded wire made of 7 strands of 0.017mm wire. This yields a cross-sectional area equal to a 1mm wide wire.

Solid wire and stranded wire both use the same AWG standard. The difference is that a certain number of smaller wires must be spun together to make a stranded wire of the same diameter as a solid one of the same gauge. Refer to **Figure 1-1** and you will see two wires that are actually the same size, the stranded wire just looks larger because its strands have been spread out. Incidentally, this flaring of the wires is called **bird-caging** and is generally not desirable.

Table 1-2 also lists the area of the wire in cmils (circular mils). This is merely the diameter of the wire in mils squared or $mils^2 = cmils$. A 24 gauge wire has a diameter of 20.1 mils so the area in cmils would be 20.1^2 = 20.1 X 20.1 = 404.01. Listed next is the resistance for a 1000 foot length of that wire gauge. A thicker wire, because of its increased conducting area, would have less resistance than a thin wire.

The maximum current rating is of importance in all circuit designs. The engineer selects a particular gauge because it safely meets the current requirements of the circuit. Placing a 38 AWG wire in a circuit that needs 100 mA (0.1A) is a sure way to circuit failure (in the best case), injury or death in the worst. Note that these continuous current ratings are for a wire in free air; if the wire is in an enclosed area, the current carrying capability of the wire must be decreased. This current rating is the wire's **ampacity** – the wire's ability to conduct the specified current safely. If you wish to double or halve the ampacity of a circuit, the general rule of thumb is to go up or down three wire gauges.

The last column isn't applicable to electronics. The ***Fuse Rating*** is the current at which the wire will burn apart as if it were a fuse. This is not a desirable event in electronics equipment.

Insulation

The importance of the insulation surrounding a wire is often overlooked. The purpose of insulation is: a) to prevent short circuits, b) to prevent damage to the conductor *from* the surrounding environment and, c) to prevent damage from the conductor *to* the surrounding area.

That a wire needs to be insulated to prevent short circuits is obvious. Unwanted current paths are a hazard to man and machine. Current is the primary factor in determining the severity of a shock. The threshold of sensitivity to current in most people is about 1mA (0.001A). However, even very low currents, on the order of 7mA
(0.007A) can be fatal.

Excessive voltage will break down an insulator and will make short circuits a possibility. It is because of this that insulators have a maximum operating voltage rating. Additionally, many have ratings for temperature, fluid resistance, fume release (toxicity), abrasion resistance, and flammability. Many off-the-shelf hook-up wires of 20 to 24 AWG might have insulation that is not oil proof, is limited to environmental temperatures not to exceed 60°C (140°F), and can't handle voltages greater than 100V. Make sure you know the operational and environmental conditions before you select a wire (whether it is solid or stranded type) and the insulating material sufficient to safely do the job.

The most common insulators used in electronics industry are made of polyvinyl chloride (PVC), rubber, or Teflon®. Quite often, these materials are mixed with other compounds to achieve a desired set of characteristics.

Kapton®, Kynar®, Tefzel®, and Poly-X® are a few of the less common insulators. They found much of their initial application in the airline industry. Their usage has ceased there due to failure to meet new FAA requirements but they have found new life in general electronics. Wire wound with fiberglass is still used in aviation but it can be troublesome to work with. Wound cotton was employed as an insulator in commercial electronics and vehicles from the 1930's to the 1950's. As you might expect, cotton has a tendency to unravel and is susceptible to chemical degradation. Nylon also found uses as an insulator but it too has fallen into disfavor.

Kapton	Very costly, good insulator, light weight, high ignition temperature, water permeable over time, ignites only at very high temperature but explodes when ignited
Kynar	Costly, poor insulator, low permeability to liquids and gasses
Teflon (TFE)	Very costly, fair insulator, physically weak but chemically resistant
Polyethylene	Lacks flexibility compared to PVC, low cost, good insulator
PVC or Vinyl	Flexible and soft, inexpensive, fair insulator, brittle at low temperatures, heavy, burns easily when ignited (1)

(1)UL (Underwriters Laboratories) 1007 is standard for PVC insulated hook-up wire. It specifies an 80 degree centigrade minimum capability and 300V rating

Table 1-4

PN ####

26 AWG 7/34 SPC .010 (.25MM) TFE

1000 FT (305M)

UL 1213 Use: In Office Appliances

Where Exposed To Oil At A

Temperature Not Exceeding 60 C

And Where Not Subject To MECHANICAL

ABUSE

Figure 1-3

The table above provides a general view of the characteristics of just a few insulators. Specific qualities can vary. For general applications, wires that meet standards such as the UL1007, UL1213 specifications or a Mil Spec (Military Specification), are normally safe to use. A data sheet for wires and insulators used in exotic environments is a must. Another option is to check the spool that the hook-up wire came from; quite a lot of information can be found there.

This sample label, **Figure 1-3**, tells us we are dealing with a 26AWG stranded wire (the 7/34

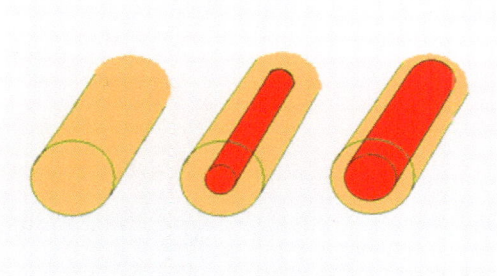

Figure 1-4
Skin effect in a solid wire. As frequency increases, the eddy current (red area) also increases.

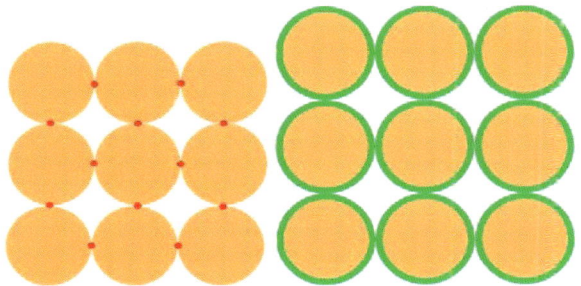

Figure 1-5
The wires to the left show proximity effect (red dots). To the right, individual strands are insulated with enamel, polyurethane, silk, etc. to eliminate proximity effect

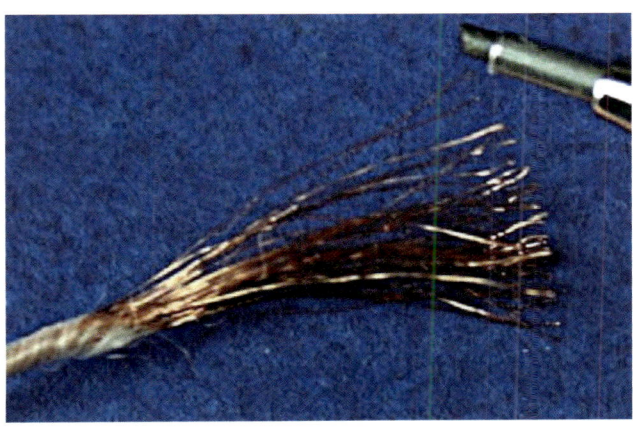

Figure 1-6
A section of nylon covered litz wire. Each strand is individually enameled.

indicates seven strands of 34 AWG wire). The insulation is 0.010" (0.25mm) of TFE (Teflon®); the wire can be used around oil but should not be exposed to temperatures above 140°F (60°C). Occasionally bits of this information can be found stamped on the wire itself. The information normally includes the maximum voltage and temperature ratings.

Skin Affect and Proximity Affect

Any time a current flows through a wire, it will take the path of least resistance. This tends to be the center of the conductor. Also, every current has an associated magnetic field. The problem is that the current leads (is developed *before*) magnetic field develops, it (current) changes a different time than the magnetic field surrounding it. The magnetic field creates currents (when the field collapses in a AC circuit)in the wires core that oppose the current that should flow through the wire (again, because of the time disparity) – these counter currents are called ***Eddy Currents***. The Eddy currents flow in the most conductive areas of the wire, its core. This isn't a problem for DCV or low frequency ACV's. However, as the frequency increases, this timing difference tends to force the current to the surface of the wire because these counter currents occupy the core of the wire. Eventually, the only path left to the AC current is the surface or skin of the wire: hence, the ***Skin Effect***. **Figure 1-4** illustrates this. The first sample shows a conductor with a DC voltage – current flows throughout the available conductor's area. As the frequency increases, the center of the wire loses its conductance because of Eddy Currents; eventually the only area left for current movement is the wire's surface.

Stranded wire has an advantage over solid wire when it comes to skin effect. It tends to be a better conductor at higher frequencies because it has many strands; the surface area is multiplied, which lessens skin effect to some degree. However, the strands introduce a new problem, the **Proximity Effect**. Any point at which two wires contact each other creates a more conductive area which causes Eddy Currents to migrate there. The way to overcome proximity effect is through the use of Litzendraht wire (commonly called Litz wire). Litz wire uses many strands of very fine wire, each of which is individually insulated. The insulator can be enamel (as in **Figure 1-6**), polyurethane, nylon silk, etc. The individual strands in Litz wire are very fine, often 36 AWG to 40 AWG. To achieve a useful current carrying capability, it is necessary to bundle many of these strands together. A 20AWG Litz wire might have as many as 41 strands of 36AWG wire each of which is individually insulated. In this image **(Figure 1-6)**; the pencil lead is 1/28th of an inch wide. This Litz wire has more than 50 strands of 40AWG.

Other Conductors

Waveguides

Even Litz wire also becomes ineffective when frequencies become high enough. At frequencies between 100MHz and 300GHz, a **waveguide** is the most efficient conductor of electrical energy. A waveguides shortfall; they can be quite large in comparison to wire or other two conductor cables. The waveguide can be a rectangular, elliptical, or circular hollow length of pipe. **Figure 1-7** is a drawing of a typical rectangular waveguide. The waveguide will vary considerably in size. Its internal dimensions are determine by the wavelength of the electrical signal and is calculated by

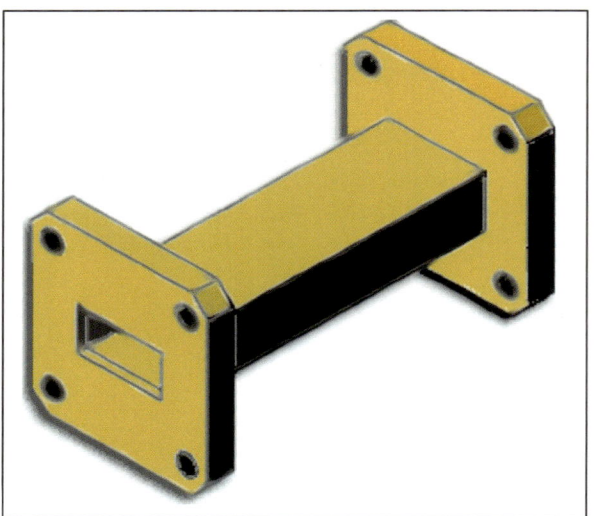

Figure 1-7

$$\lambda = 3 \, X10^{10}$$

where $3X10^{10}$ is the speed of light in centimeters per second and F is the frequency of the signal. λ is the lowercase Greek letter lambda. The answer to the equation is in centimeters. This calculation does not take many factors into account and gives a very generalized idea of waveguide dimensions. The equation tells us that the higher the frequency, the smaller the waveguide has to be. A waveguide used from 300MHz to 500MHz would have approximate dimensions of (A) 23" (58.4cm) by (B) 11.5" (29.2cm). A guide used at 100GHz would be approximately (A) 0.100" (2.54mm) to (B) 0.05" (1.27mm). These are *inside dimensions*.

You have probably heard of one waveguide, fiber optic cables. A fiber optic cable is a "light tube" (just as waveguides are "electrical tubes") and is so small because of the very high frequency of light. The frequency of light is roughly 10^{15} Hertz or 1 Petahertz.

Coaxial Cable

Coaxial cable is unlike the other conductors we have covered. Wires and waveguides are all single conductor devices: coaxial cable (often called coax or RF (radio frequency) cable) is composed of *two conductors*, a center conductor and a shield wire. The center conductor is the signal source and the shield acts as a ground. Coax is essentially an electrical cord for radio frequency signals. The radio frequency spectrum occupies the frequency band from 30,000 Hz (30KHz) to several GHz.

Developed in 1929, the cable found no initial applications. It was 1940 before AT&T used it for phone systems. Today, coax cable is found in almost every home that has a television.

Coax is used because its AC resistance (impedance) is constant throughout the length of the cable. The AC resistance is the same no matter where the cable is cut. Impedance is a concept beyond the scope of this text but a brief explanation is necessary. AC resistance is determined by inductance and capacitance. The inductance and capacitance are held constant by coaxial cables. It is the inner insulator that makes this possible. The distance that the inner insulator separates the inner conductor and shield establishes the impedance. A constant impedance is desirable because of impedance matching. Match the impedance of a signal source (transmitter) and a signal carrier (the coaxial cable) and maximum power can be transferred between the two.

We cover coax cable because of its ubiquitous nature. The impedance of a cable is determined by the distance between the center conductor and shield, not the length of the cable. The impedance can vary if the cable is crushed (which changes the distance between the shield and center conductor). t is far more likely a bad connection because of poor installation practices or deterioration of the connection from environmental factors. A cold solder connection or sloppy crimping are far more likely to cause impedance mismatches.

Coaxial cables come in many types. The **RG-58** has a 50-ohm impedance and is used in many transmitters and receivers. **RG-59** is a 72-ohm cable used in video. If the shield is a solid piece of copper, not like the braided shield seen in **Figure 1-8**, the cable is called **Semi-Rigid**. Should the cable be made of one tube of copper suspended in another tube of copper, it is called **Hard Line**.

For every type of coaxial cable, a connector must exist. We will concentrate on the Bayonet Neill-Concelamn (BNC) style of connector. A BNC connector is seen in **Figure 1-9**. It is the connection found on most electronic test equipment used below 500MHz. Chapter 2 discusses attaching a coaxial conductor in depth.

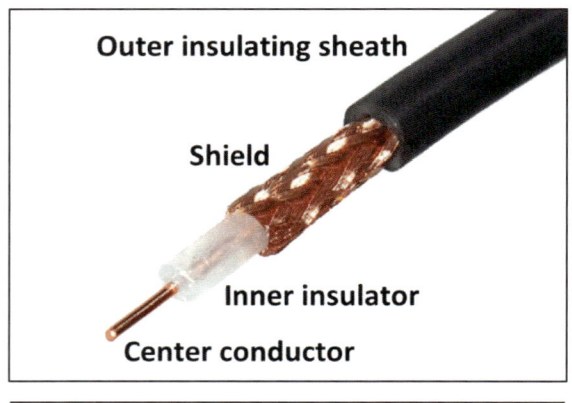

Figure 1-8

Figure 1-9

Chapter Two

Standards: Wires and Splices (and Some Cabling Too)

Every organization you work for will have job standards. This is true even of a hobbyist, his work has to meet a certain level of acceptability (his own and his peers). There are standards for something as mundane (and apparently simple) as stripping the insulation from wire. This chapter introduces you to good practices in stripping wires and various wire and cable connecting techniques.

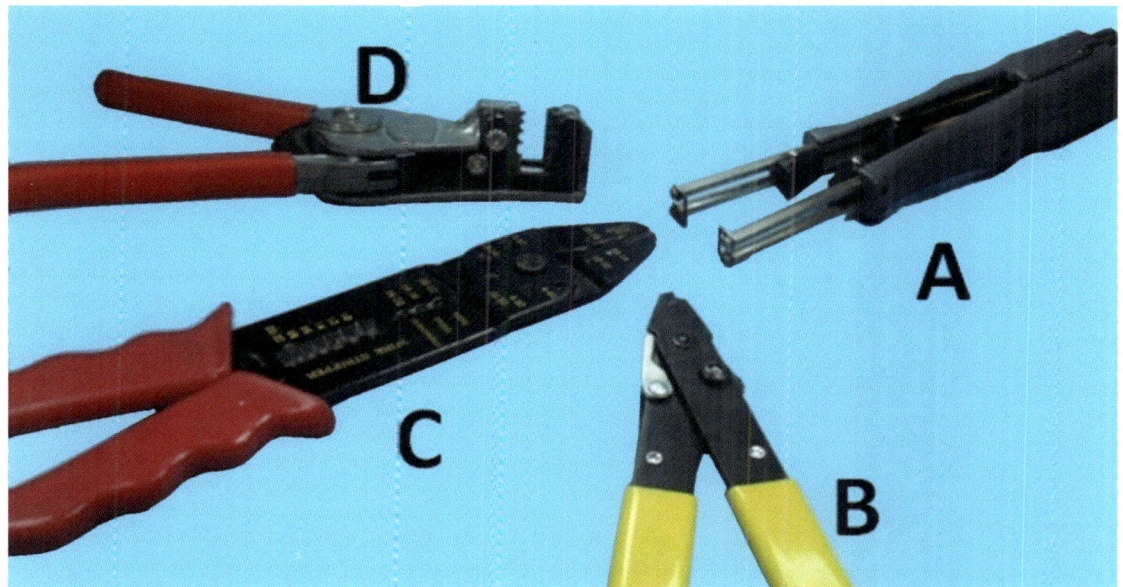

Figure 2-1

Common tools used in stripping wires. Clockwise beginning in the top-right; (A) thermal wire strippers, (B) basic manual wire stripper, (C) manual wire stripper and crimper with gauge index, and an (D) automatic wire stripper.

Wire Stripping

The idea behind wire stripping is to cut the insulation but not the wire underneath. **Figure 2-1** shows four types of wire stripping tools commonly used to perform this task. Each tool has advantages and shortcomings. One tool you may have never seen is the thermal wire stripper in the top right (A) of the photo. It uses heat to melt the insulation, which makes it almost impossible to damage the conductor. Thermal strippers can be slow as it takes a few seconds for the tools tips to heat and melt the insulation. Additionally, such strippers normally have a temperature adjustment must have the temperature set properly for the type of insulation to be stripped. PVC and rubber insulation need a

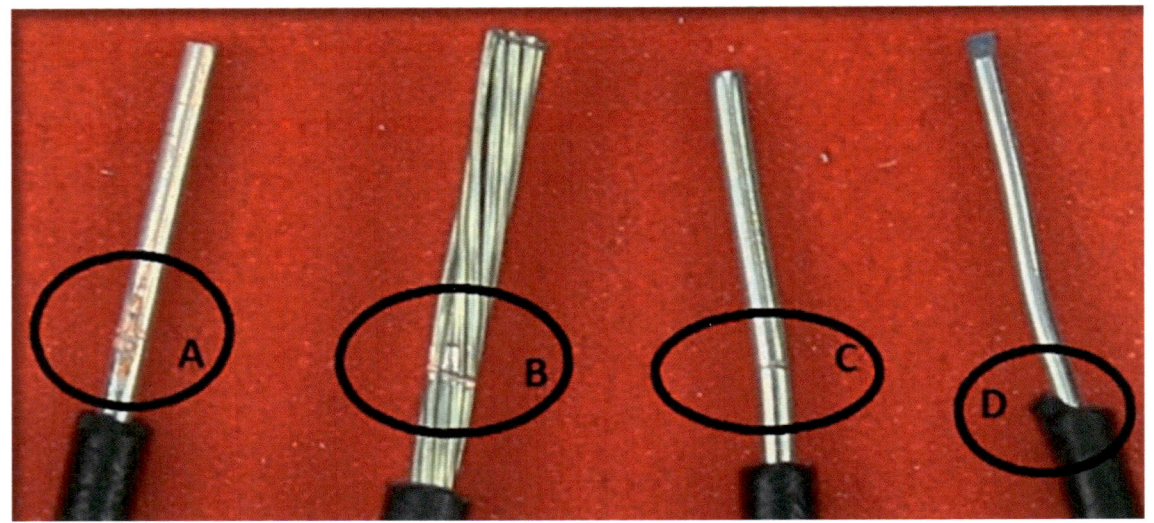

Figure 2-2
This image shows four types of damage that can occur on wires or insulators.

lower temperature than Teflon™ insulation. The manual wire stripper (B) works much like a pair of scissors but has a notch in which the wire is placed. A certain amount of practice is required with this tool to prevent nicking or scraping of the wire. Once mastered, this device strips insulation very quickly. The manual wire stripper with the gage index (C) is somewhat better than the previous item. Wire can still be damaged if the technician doesn't pull the wire perpendicular to the tool or uses the wrong gauge slot. The automatic wire stripper (D) has calibrated cutting teeth to remove the insulation without damaging the wire. It also draws the wire and insulation apart as it cuts.

Wires can be damaged no matter what tool is used. The wires in **Figure 2-2** exhibit various forms of damage that can be caused by incorrect tool operation. Wire A has been scraped and the factory tinning (a thin layer of solder applied to prevent corrosion and make soldering easier) has been removed. This has exposed the copper to atmosphere, which leads to oxidation. The stranded wire B has been nicked. Because stranded wires are so thin, these strands will likely break after being bent once or twice. Solid wire C has also been nicked. This type of nicking can be invisible until the remaining insulation is pulled back. The nick occurred where the insulation was cut. All three examples (A, B, and C) lead to mechanical weakness in the wire and a point for contaminants to deposit. Example D shows insulation that was not cut cleanly. All such irregularities offer areas for contaminants to attach and for corrosion to set in.

Number of strands	Less than 7	7-15	16-25	26-40	41-60	61-120	Greater than 120
Class 1 and 2 Maximum number of strands nicked, scraped, and missing	0	1	3	4	5	6	6%
Class 3 Maximum number of strands nicked, scraped, and missing for *tinned wire*	0	1	2	3	4	5	5%
Class 3 Maximum number of strands nicked, scraped, and missing for *wire not to be tinned*	0	0	0	3	4	5	5%

Table 1

Table 1 defines the limit of damage in a stranded wire. Class 1 and 2 are most forgiving. The Class 3 standard for *wire that is not to be tinned* has the strictest criteria.

How to Test For Proper Wire Stripping

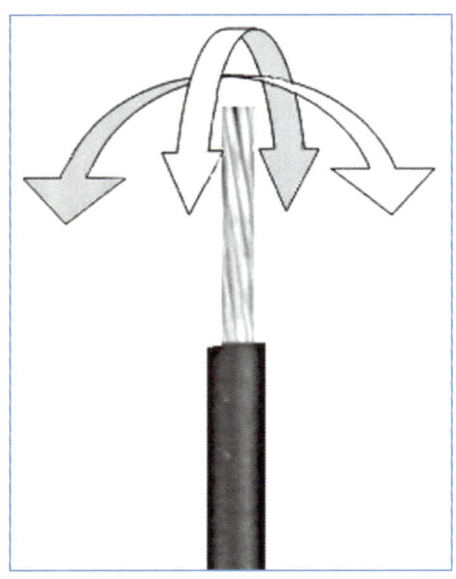

Once a wire has been stripped, it should be in perfect physical condition. Flaws in solid wires are usually easy to find. Visual inspection generally suffices. To make sure a problem is not overlooked, pull the insulation back on any solid wire after you have removed the insulation. You don't need to do this every time, just while learning. This will reveal the area that is most often damaged but rarely seen. The point where the tool cuts the insulation hides these flaws; what might appear perfect is not necessarily so. If you knick a wire, you will be decreasing its circular area (making it less conductive) and leaving a mechanically weak area.

Figure 2-3
This sample image shows the original wire twist and the directions for bending the wire when testing.

Stranded wire can be a little more troublesome to check. When you remove the insulation, the strands should lay as they did before you removed the insulator; they should retain their original twist (see **Figure 2-3**). Wire that flairs is said to have **_Bird-Caged_** (see the stranded wire in **Figure 1-11** for an example of this). It is often difficult to see damage to the small strands. One way to test your work, during the learning process, is to take the wire and bend it back and forth seven or eight times (hold the wire just below the point you just stripped). Repeat the procedure by bending the wire perpendicular to the first set of bends. A properly stripped wire will not lose any strands.

Wire Connections

The most basic electrical connection (or repair) you will probably make is wire to wire. Splices can be classed into two categories. **Butt splices** are any join in which wires are connected end to end. **Tap splices** are those in which one wire is inserted into a preexisting run. There are many methods for making a butt or tap splice, the eight methods shown here are only representative. The first six – appliance splice, Western Union splice, tap splice, mesh splice, pigtail, and lap splice are all solderable connections.

The Appliance Splice

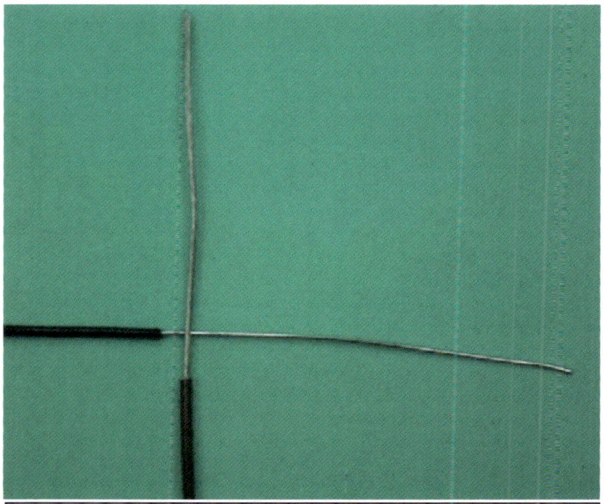

The appliance splice is somewhat of a rarity today but it is useful in illustrating some important points. The appliance splice is used to connect a solid wire to a stranded wire.

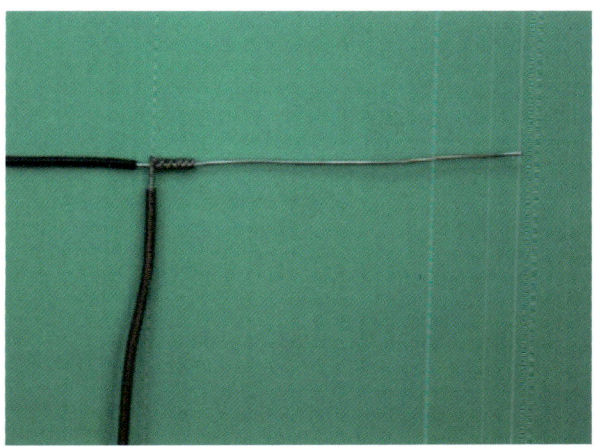

Figure 2-4a

After stripping about 2.5" (6 cm) of insulation from each wire, place the stranded wire over the solid as in **Figure 2-4a**.

Tightly wrap the stranded wire around the solid (**Figure 2-4b**). This is done because the stranded wire is the more flexible of the two types. Ensure the twist of the stranded wire remains the same as it was while the insulation was present.

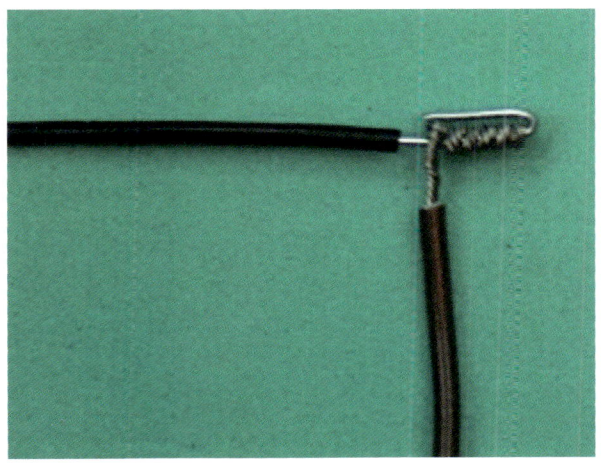

Figure 2-4b

Once you have the stranded wire wrapped five times around the solid, cut the stranded wire and bend the solid wire back along its length. Cut the solid wire even with the first stranded twist (**Figure 2-4c**).

Because this splice and those that follow can be soldered, allowance must be made for a ***Solder Gap.*** A solder gap is an un-insulated length of wire between the insulation and the connection of another wire/component, which prevent capillary action from drawing solder under the insulation during the soldering process. In wire connections, generally one to two insulation widths are acceptable. Take piece insulation from the same wire used in making the attachment. Place it into the gap between the insulation and the beginning of the work. The distance must be at least one solder gap (one time the width of the insulation)

Figure2-4c

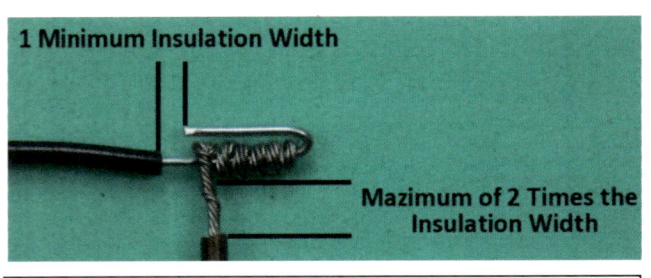

1 Minimum Insulation Width

Mazimum of 2 Times the Insulation Width

Figure 2-4d

but never more than two solder gaps (two times the width of the insulation). See **Figure 2-4d**. The appliance splice is forgiving because its gaps can be adjusted. The splices that follow do not allow for these adjustments.

The Western Union Splice

The Western Union Splice goes back to the days of the telegraph. The splice was designed to support heavy wires strung over long distance. It is mechanically strong. Once it is soldered, the splice has more mechanical strength than the original wire. This technique can be used for solid and stranded wire.

Begin by crossing the wires to be connected. If using stranded wire, make sure the wires have the same twist as they did under the insulation. Notice in **Figure 2-5a** that the wires are offset. The intention is to twist both wires simultaneously about an axis. Since you are applying this twist with only one hand (the wires in the figure are set to be twisted with the left hand), leave extra space on the side you are twisting. You do not have to grip the wires firmly. Your fingers are acting as guides during the motion. Always twist around the axis and make sure there is always an angle throughout the twisting motion. If you fail to do this, the appearance of the wires becomes mushy and indistinct. When done correctly, the wire should look

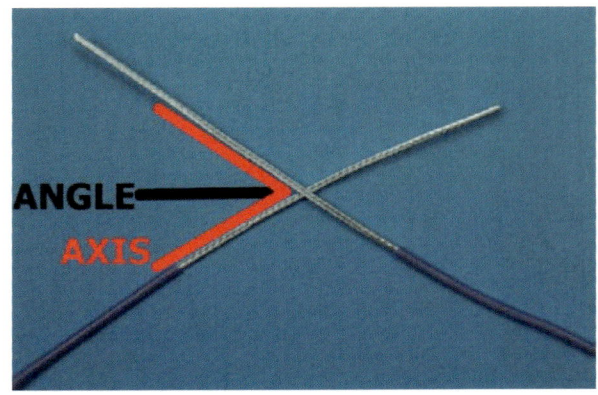

ANGLE
AXIS

Figure 2-5a

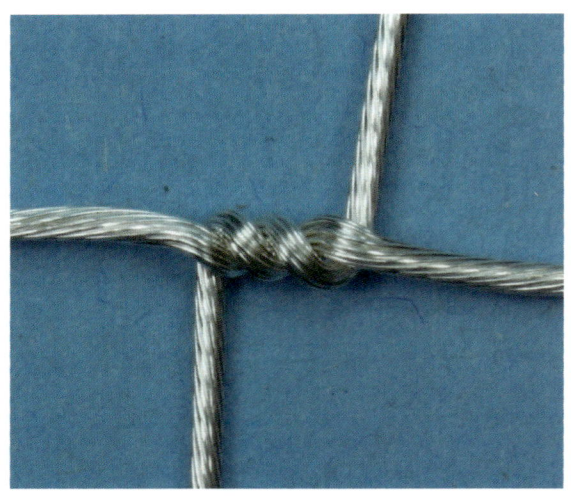

Figure 2-5b

like **Figure 2-5b**. Notice that the twists are readily visible. Get three twists centered on the wire. Once this is done, adjust the wires ends so they are perpendicular to the direction of the wire run.

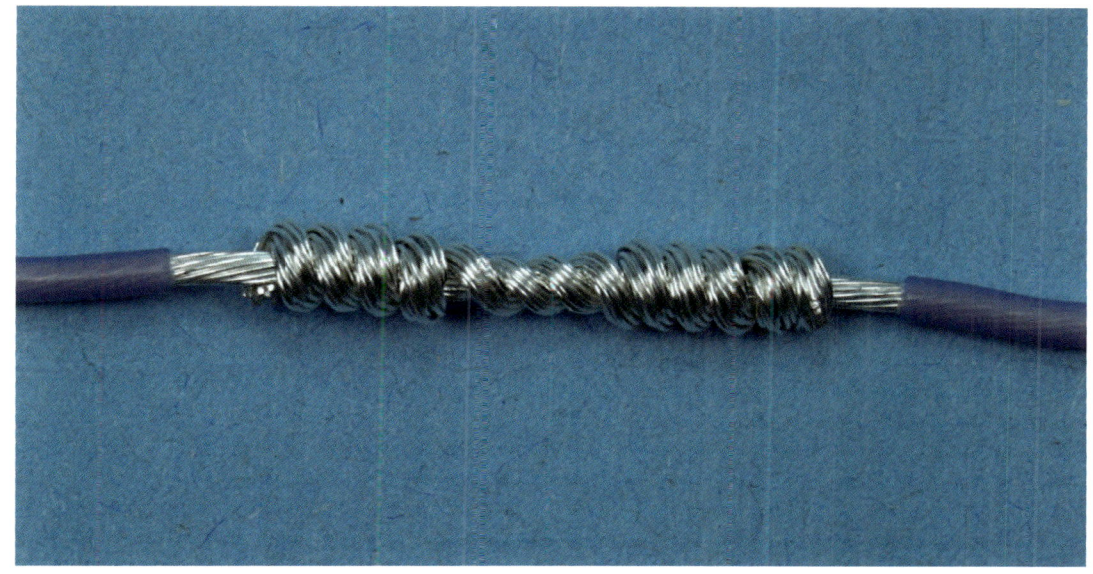

Figure 2-5c

Figure 2-6

Once the center twists are done, wrap the loose wire around the center conductor five times. Cut any excess wire. When the Western Union splice has been completed, it should look like **Figure 2-5c.** The original wire twist is still visible and there are no gaps between any of the wraps. The ends of the wires should not stick out because once soldered, these ends could penetrate any material you use as an insulator such as electrical tape (this is not the kind of electrical tape found at the local discount store) or heat shrink Make sure none of the strands are damaged and the insulation is cut cleanly. Finally, remember to leave a solder gap of one insulation width minimum, two widths maximum on each end.

The Western Union splice will take some practice to get right. The length of the wire you must strip and where to begin twisting the two together will vary by the wire's gage. Make mistakes and your work will look like **Figure 2-6**. This photo exhibits most of the mistakes that can be made. The solder gaps are too large, the ends of the wire aren't flat and have bird caged, the strands aren't in their original twist, and the turns of the wire are not defined.

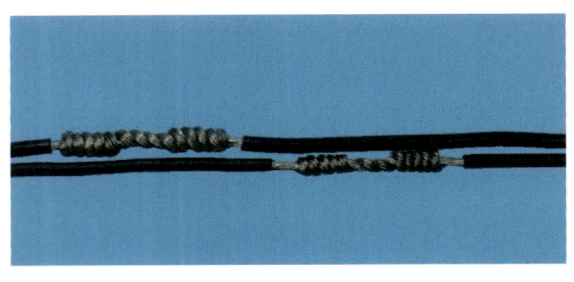

Figure 2-7
Staggered splices

Occasionally several splices must be made in the same area. If this is the case, make sure to stagger the connections (see **Figure 2-7**). The splice is often thicker than the original wire. Forming all the splices at one point creates a bulge that is not only unattractive but also difficult to work with. Above all, if the splices are in areas subject to movement, they could chafe and short circuit or open.

The Pig Tail

The pigtail splice is another type of butt splice. It lacks the mechanical strength of the tap splice and must be soldered before it is bent into its final position. It makes up for these limitations with its simplicity of construction. The three steps to construct a pigtail are shown in **Figure 2-8a** through **Figure 2-8c**.

To make the pigtail, simply cut two wires and strip equal lengths of insulation, twist five times, clip the excess, and fold over. Solder the connection before folding the wire over.

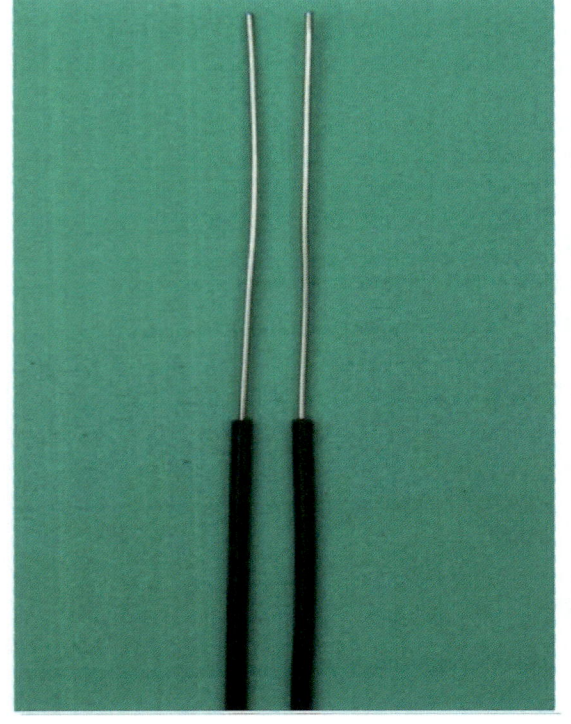

Figure 2-8a

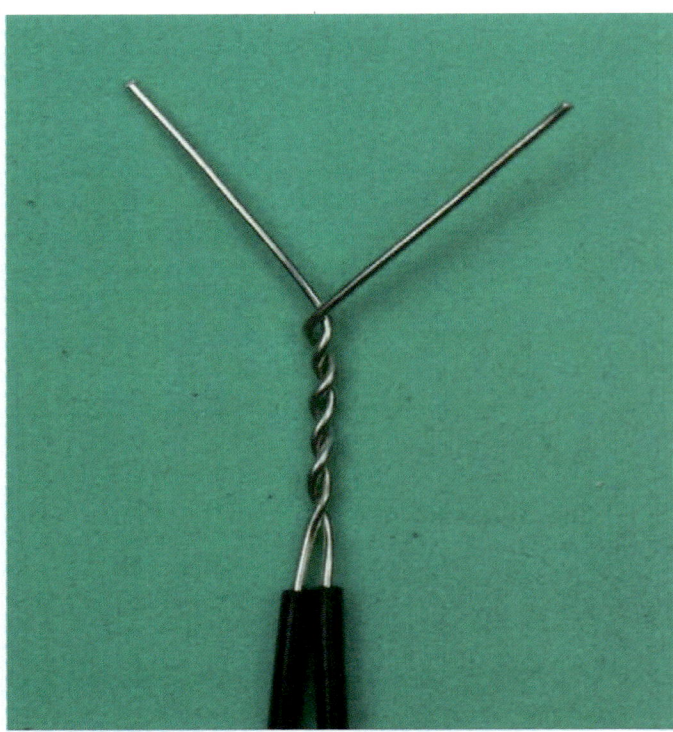

Figure 2-8b

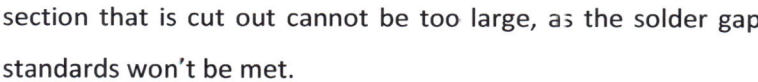

Figure 2-8c

The Tap Splice

The tap splice is used to make a new connection into an existing wire run. It can be a challenging task to complete properly. A piece of insulation must be cut out of the wire into which you are splicing without doing any damage to the existing wire. Additionally, the section that is cut out cannot be too large, as the solder gap standards won't be met.

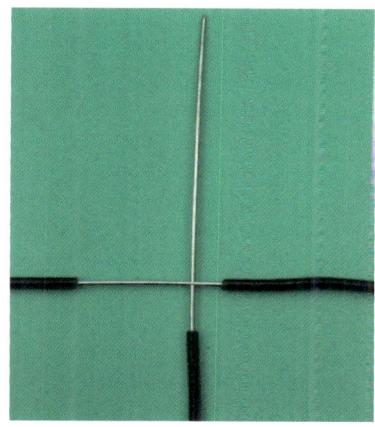

Figure 2-9a

This series of images shows the process of making a tap splice. Again, cutting out the correct amount is critical. As a rule of thumb, it is always better to cut out too little than to cut out too much. You can always remove more insulation but you cannot put insulation back!

Once the wire is wrapped around the back, bend the loose end left or right (**Figure 2-9c**) and bend the loose end towards you. Securely wrap it around the tapped wire. You should be able to get five wraps around the tapped wire and maintain the solder gaps. You should not see any gaps between the loops, the end of the wire must be down and the insulation must be cut cleanly.

Figure 2-9b

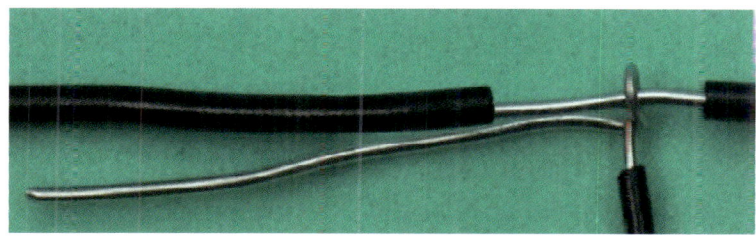

Figure 2-9c

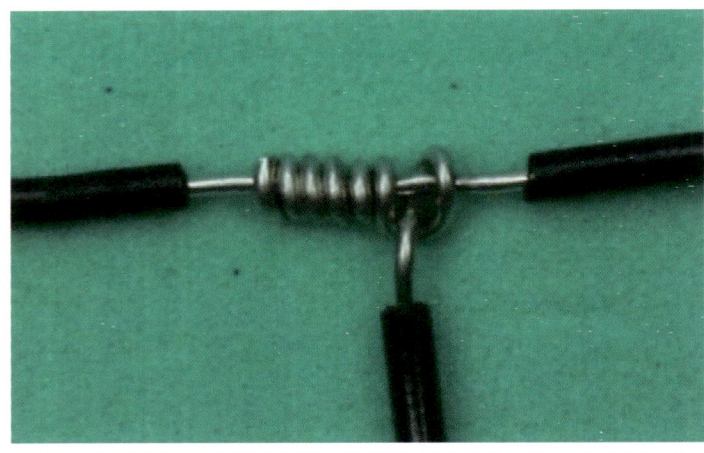

Figure 2-9d

A properly made tap splice should have this general appearance. Note the solder gaps are one to two insulation thicknesses wide. Your measurement must be accurate. You cannot add more wire to the wrap to get five wraps. You can remove insulation to make room for the five wraps. Be sure to maintain proper solder gaps while doing this. Heat shrink can be used on a tap splice (Chapter 4). If heat shrink is used, bend the tapped wire parallel to the original run and solder the connection.

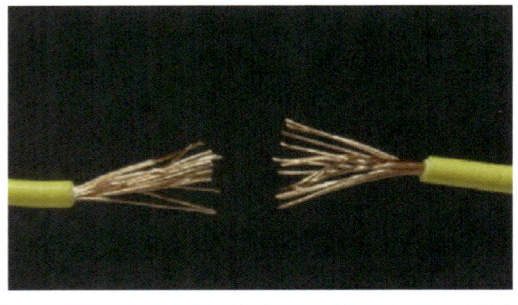

Figure 2-10a

The Mesh Splice

The mesh splice is used to join stranded wires end to end. Soldering is required for this splice (and the lap splice, which follows). You may want to read and practice the material on soldering in Chapters 3 and 4. Start by stripping at least 5/8" (16 mm) of insulation off each wire (**Figure 2-10a**). This allows the recommended overlap of ½" (12.5 mm) required of a properly made mesh splice. When the ends have been joined, flatten the wires and leave a solder gap as shown in **Figure 2-10b.** Finally, solder (**Figure 2-10c**) the connection and heat shrink to insulate and protect the splice.

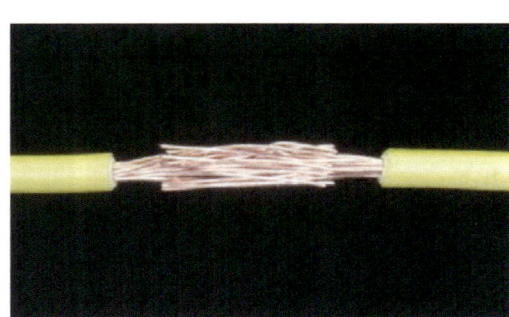

Figure 2-10b

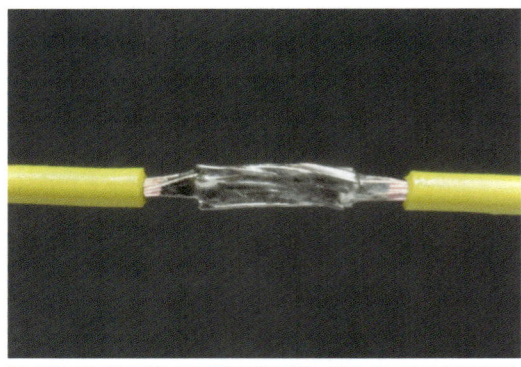

Figure 2-10c

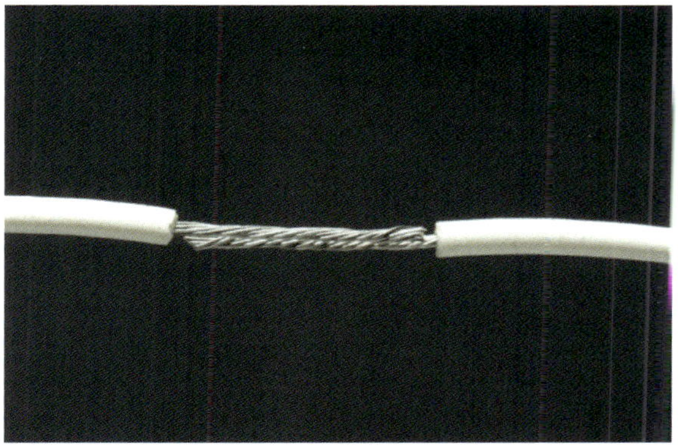

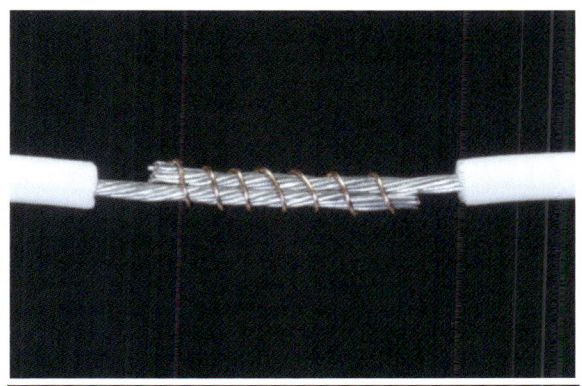

Figure 2-11a

The Lap Splice

The Lap splice will work for solid and stranded wire. Remove insulation roughly equivalent to 6-8 times the insulation thickness. A lap splice must over lap at least four times the insulation thickness (**Figure 2-11a**). When the wires are stripped, and placed together, wrap a 30 AWG wire around both to secure them (**Figure 2-11b**). When secured, solder the overlapping wire (**Figure 2-11c**). Again, leave a correctly sized solder gap. Secure the connection with heat shrink.

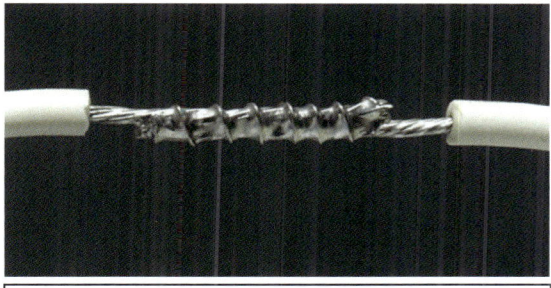

Figure 2-11c

Figure 2-11b

Crimping Tools and Splicing

Quality splices can be achieved using crimping tools. However, the quality of a splice is now more dependent on tools and supplies than on your skills alone. The first critical element is the type of splice you are going to use. The image (**Figure 2-12**) below shows two common butt splices. The one on the left is a high-grade splice suitable for use in aviation. The splice shown to the right should be avoided if possible. The arrow points to a small ridge on the quality splice. The crimping tool will compress this ridge and force it onto the insulation, holding the crimp securely in place. In this way, the insulation becomes a lock for the crimp, not the wires strand(s). The portion of the bare wire

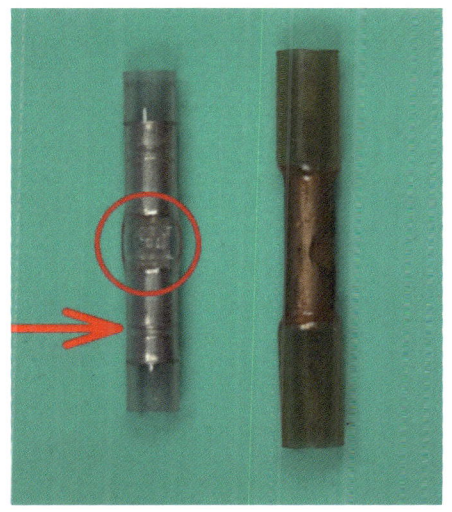

Figure 2-12

under the crimp is only compressed lightly; this eliminates damage to the delicate strands. The crimp to the right in **Figure 2-12** requires that it be compressed onto the strands, which will definitely damage them. This crimp also lacks a way of assuring mechanical alignment when placed in a crimping tool. The circled area in the good crimp (left of **Figure 2-12)** is a notch used to align the crimp/splice in the crimping tool.

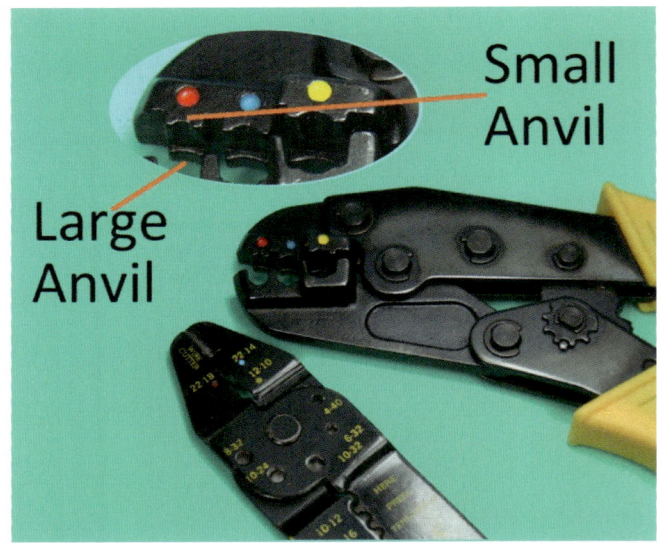

Figure 2-13
Two common crimping tools. The tool to the right is the preferred device for proper crimping.

The second important element to achieving a proper crimp requires a quality tool. Two crimping tools are shown in **Figure 2-13**. The tool to the lower left can be purchased almost anywhere but cannot make a high quality connection. The other tool, however, has a set of dies calibrated and color coded to the proper splices. Both of the splices in **Figure 2-12** are red so the red dotted part of the die would be used. Splices are color coded to denote which wire gauge they will work on. As a rule, red crimps are for 18-22AWG wire, blue for 14-16AWG, and yellow for 10-12AWG.

You will notice there are two anvils under each dot (shown in the enlarged section). These are used to crimp both the insulation and the wire to the proper depth in one-step. You can see in **Figure 2-14** how the tool locks the crimp in place (the rearward large anvil)

Figure 2-14

and how the small anvil in front compresses the crimp on the insulation and wire. **Figure 2-15** shows what such a crimped splice would look like. Notice the shallower crimping on the insulation, heavier crimping on the section making contact with the stripped wire.

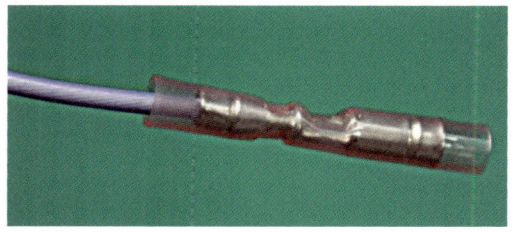

Figure 2-15
A completed crimp. Make sure the wire was placed deeply enough into the crimp so the ring can lock onto the insulation. not the conductor.

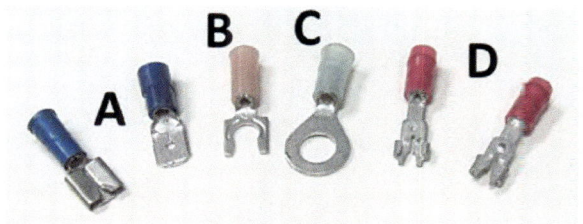

Figure 2-16
(A) female (sometimes insulated) and male blade disconnects, (B) fork or spade terminal, (C) ring terminal, and (D) locking fork terminals.

Solder Splicing Sleeves

Another connection technology is the solder splice. This is a splice which has a heat shrinkable body and a solder ring in the center of the connector. When a heat gun is used, the heat shrink contracts around the wire. As the temperature increases, the solder reaches its melting point and flows into the exposed wire. The red rings on either end are a sealing sleeve which also contracts as it is heated. The rings are also the gauge indicator for the splice

This splice is very effective when properly employed. There are some words of caution however. The most important point - make sure that the wire's insulation you are splicing can withstand temperatures required to melt solder and that the

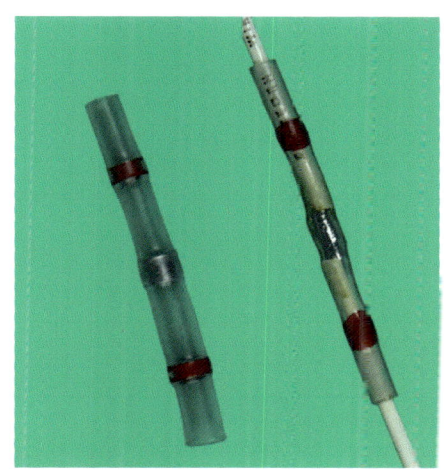

Figure 2-17
A solder splicing sleeve before and after use.

connection is not used at 125°C or higher temperatures. This connection cannot be cleaned once it is made. As with all solder, this connection can reflow if heated sufficiently. Clean the wires thoroughly before permanently joining them. If you must use flux, make sure it is a no-clean type.

RF Cable Connectors

Attaching a BNC connector to a cable can be done with simple and relatively inexpensive tools. **Figure 2-18** shows some of the styles available. The upper and lower crimpers will work on two sizes of BNC connector, those designed for RG-58 and RG-59 BNC cable connectors. The crimping area of tools A and C, the width of the crimping jaws, is narrow compared to the crimping sleeve (**Figure 2-20**). It s

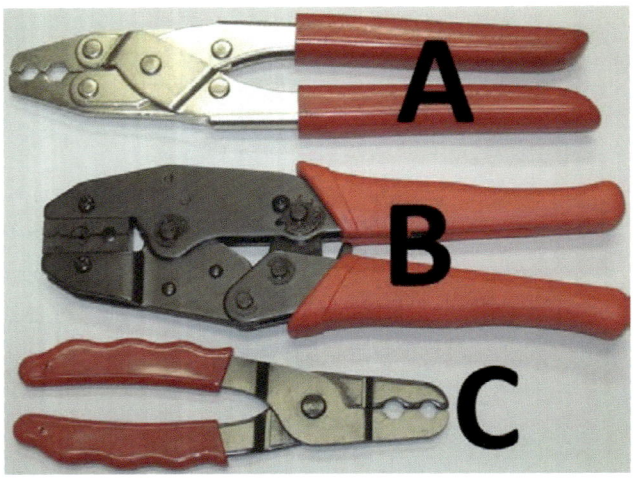

Figure 2-18

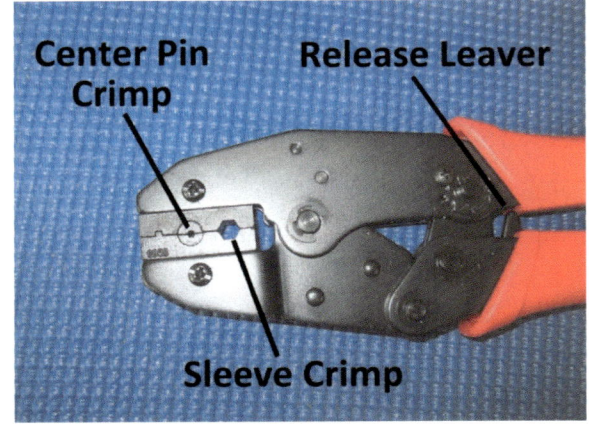

Figure 2-19

necessary to crimp the sleeve several times to make a proper connection. Additionally, these two crimpers are not designed to crimp center pins. Crimper B has wide jaws that will crimp the entire sleeve at once and can also crimp a center pin. The tool shown has jaws specifically for connectors used on RG-58 cables. Jaws for other sizes can be installed. If you make a mistake with the crimpers (crimper locked with the sleeve misaligned) in the center of the figure, you will have to release a gear to get them to open. As you close the tool, a ratcheting gear locks the jaws. To open the jaws, press the leaver (shown in **Figure 2-19**) forward and the tool will spring open. The slots for the sleeve and center pin crimping slots are also seen.

A common RG-58 BNC connector is seen in **Figure 2-20**. The center pin is to the left, the sleeve to the right. A small hole in the center pin indicates that this pin can be soldered onto the center wire.

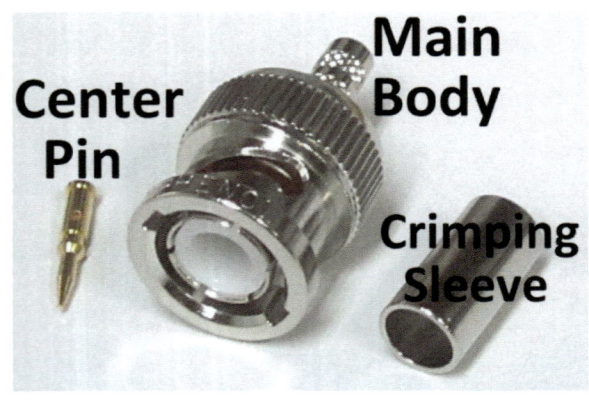

Figure 2-20

Once the type of cable and connector have been decided on, you will need to know what length the sheath, shield, inner insulator and center conductor will need to be cut to. For the connector in **Figure 2-20**, the manufacturer of the connector specifies these dimensions:

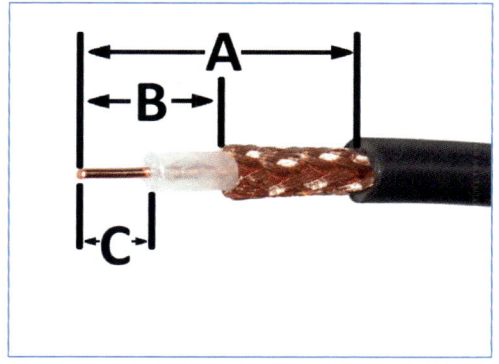

Figure 2-21

Stripping Dimensions, inches (mm)

A = 0.593 (15.1)

B = 0.250 (6.4)

C = 0.156 (4.0)

Figure 2-21 indicates where these dimensions are applied. Strip the outer sheath to dimension A using either a coax cable stripper (sized for the cable) or some other suitable tool. Do not cut any wires of the shield. This is one of the few times a knife can be effectively used for wire stripping. Lightly cut the sheath with the knife but *do not* get close to the shield wire. Now, bend the sheath along the lines you scored and the sheath will separate at the cuts. The shield will now be exposed. You can remove about 1/5" (5 mm) of the braided wire. More can removed but it is always prudent to take a little less off initially. Remember the adage, "You can always remove more but you can't *add* more". Place the crimp sleeve on the cable now. Push the shield back as indicated by **Figure 2-22**. This clears the shield from the inner insulator and allows the insulator to be trimmed to dimension C. Remove the shield to length B but avoid unraveling the shield. The shield should remain woven as it was originally. At this point, solder or crimp the center pin to the center conductor. If you measured the C dimension correctly, the pin should cover the entire center connector and the pin's base should touch the center insulator. Slide the body of the BNC plug over the center insulator.

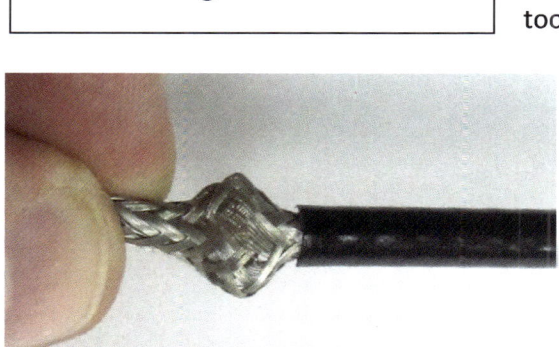

Figure 2-22

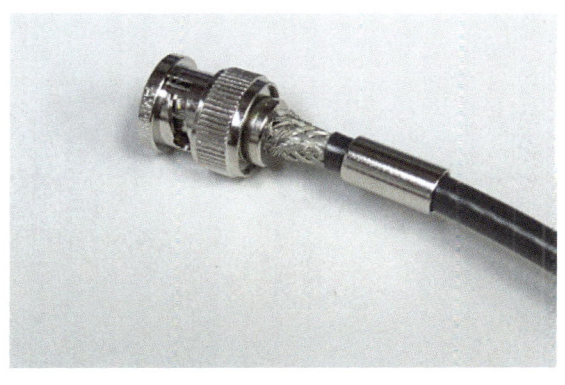

Figure 2-23

31

The shield wire should go over the base of the BNC connector. Trim the shield wire as necessary and slide the sleeve up the cable and flush to the BNC body (**Figure 2-24**). Before crimping the sleeve, be sure the *center pin* reaches to but not beyond the white insulator ring as shown in **Figure 2-25**. The sleeve has also been crimped. The connection can now be tested.

Figure 2-24	**Figure 2-25**

This is only one style of connector among the hundreds that are available. No matter the type you work with, always: 1) have the data sheet for the connector available, 2) never damage any part of the RF cable, particularly the shield and center wire, 3) do not distort the inner insulator by crimping the sleeve too hard, 4) be sure the center pin extends out sufficiently but not too far.

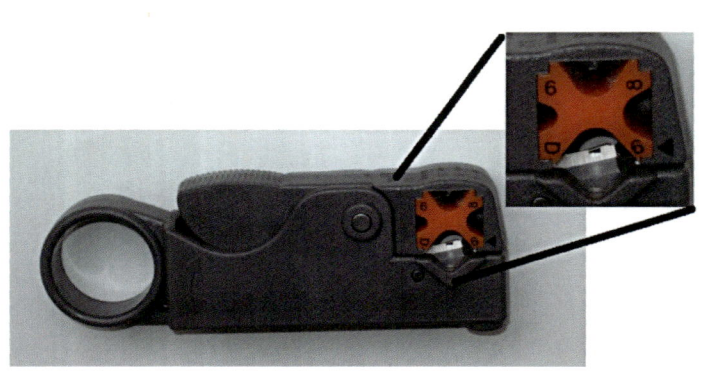

Figure 2-26
This is a commercially available RF cable stripper. You open the unit by pressing on the knurled lever, insert the cable to the stop block, and rotate in the direction indicated on the tool. The red "x" in the center is removable and is rotated to adjust the depth of cut for different cables.
This strippers' depth of cut is adjustable via two Allen screws.

Molex Connectors

The last connector we will cover is the Molex. Molex connectors, such as the one found in **Figure 2-27**, are extremely common in computers, aviation, and solar power. You may be most familiar with these connectors from their use in computer systems. All of the power connections are Molex type connectors.

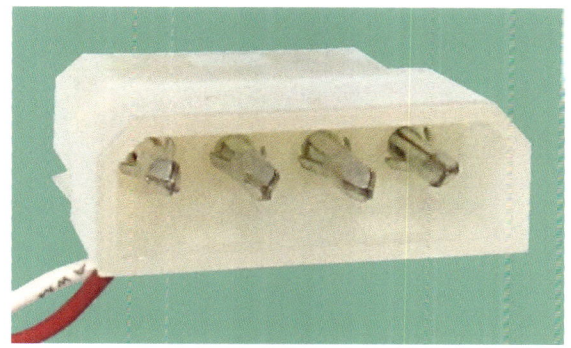

Figure 2-27

Molex connectors are popular because many circuit joins can be made quickly and the plugs can be disconnected allowing easy service or replacement of instruments and modules. The pins can also be removed using a tool that is little more than a hollow tube. This tool is just wide enough to slide over the pins. Molex pins have a pair of locking tangs (**Figure 2-28**) which keep them from sliding out of the plastic module. The tool presses the tangs inward and allow you to pull the pin out of the back of the module.

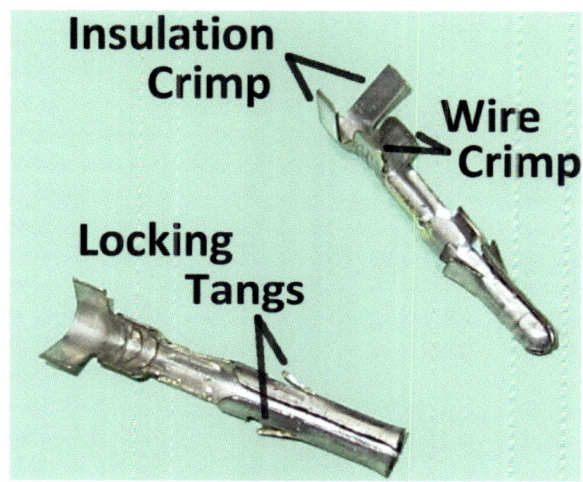

Figure 2-28
Female and male pins for use in a Molex connector.

Figure 2-29

Notice the shape of the anvils seen in the inset. Their shape properly folds the insulation and wire wings and applies the correct amount of pressure for firm contact.

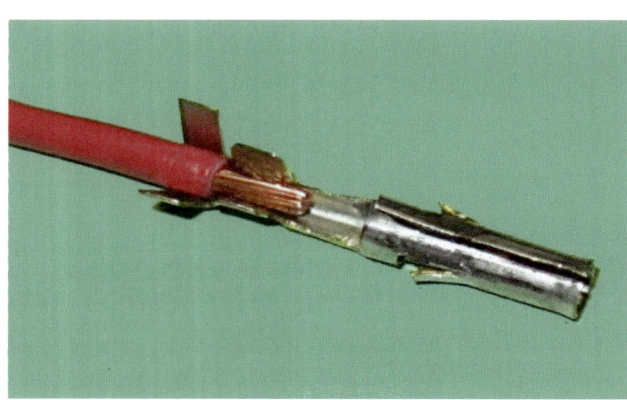

Figure 2-30

Molex connectors also have sound electrical and mechanical characteristics. They crimp onto the wire's insulation as well as the conductor. You can see the wings that will be crimped onto insulation and wire in **Figure 2-28**. A crimper designed for Molex pins must be used to connect pin to wire. Many crimp/pin sizes are available. Therefore, you must chose not only the right pin for the job but also the correct size of dies for the crimp tool. Some tools have dies that can be changed to suit the crimp. Others, such as the one in the picture, have fixed dies. Some of these tools are like a pair of modified pliers while others are a little more complex. They may have a calibrated set of anvils and a wire support bracket (on the opposite side). Both simple and complex crimpers will have two slots, often designated A and B. The B slot crimps the insulation, the A crimps the wire (**Figure 2-29**).

Strip the wire as directed by the instructions. **Figure 2-30** illustrates how the stripped wire should look before crimping. Place the crimp into slot B of the tool first and then align the stripped wire in the crimp. Close the tool. This will secure the insulation to the crimp. The crimper may be a ratcheting type. To open the jaws, you have to press the tool until it stops clicking. A spring will then open the jaws. The insulation should now be crimped. Place the wire crimping section of the pin into the B area and crimp. Now move the crimp to A and crimp the wire once again. This completes the connection. An example of a Molex pin properly crimped to a wire is shown in **Figure 2-31**. All that remains is to insert the pin into the molded housing.

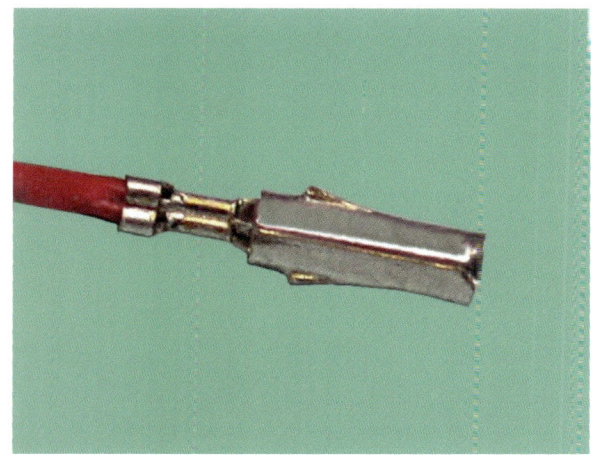

The housings, just like the pins, come in male and female styles. The housing in **Figure 2-27** is a female type. Female plugs will have male pins; male plugs have female pins. The holes in the plugs are also numbered and/or lettered. This aids in getting the correct wire into the correct slot and this *helps* prevent cross-connecting wires.

Figure 2-31

RJ45 Plugs and Jacks

The Registered Jack 45 (RJ45) is a modular connector used in wired ethernet communication systems. The RJ45 you are most familiar with is the eight pin, eight contact (8P8C) *plug* (component to left in **Figure 2-32**) that is inserted into your computer, router, or modems *jack*. The jack is the item to the right in **Figure 2-32.** This figure also shows the position of pin 1 on each part.

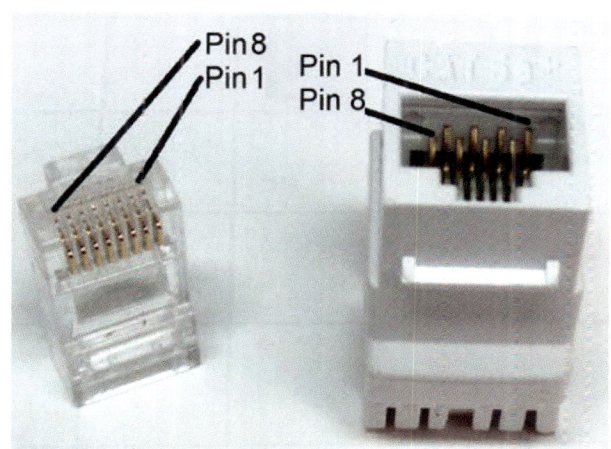

Figure 2-32

RJ45 jack (left) and plug

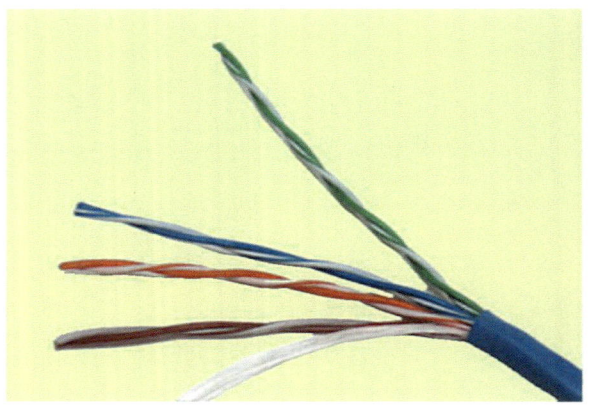

Figure 2-33

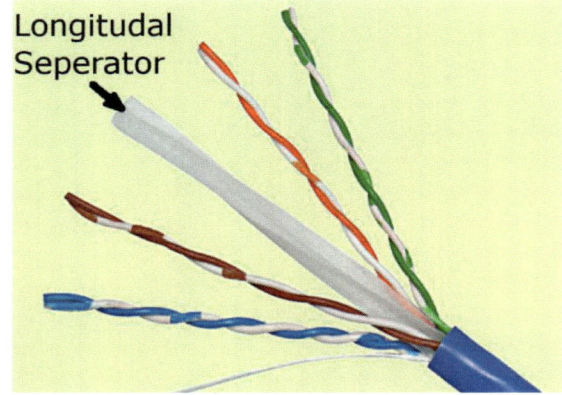

Longitudal Seperator

Figure 2-34

Two types of cable are normally used with RJ45s, the Cat5e and Cat6. Both have eight wires but the Cat6 cable features a *longitudinal separator* absent from the Cat5e. This separator keeps the wires in a specific position and is the feature which extends the frequency capability of Cat6 cables. In each cable, two wires are wrapped around one another to form a *twisted pair*. The pairs are: 1)green/green and white, 2)blue/blue and white, 3) brown/ brown and white and, 4) orange/orange and white. **Figure 2-34** is a Cat6 cable. The white strands at the bottom of **Figure 2-33 and 2-34** are thin nylon cords.

Twisted pair are used because they improve signal quality via complementary signals. One wire would transmit a normal signal, the second transmits a inverted signal – this is *complementary signaling*. The signals are sent through a device which compares the two signals. Signals of opposite amplitudes add together to make a stronger signal. Like amplitudes subtract.

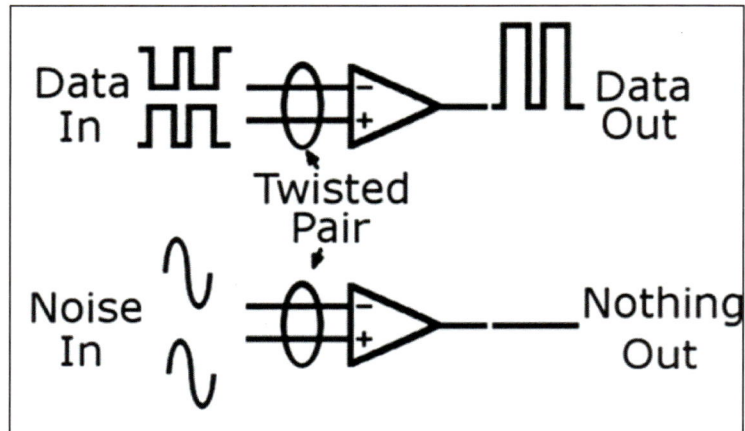

Data In — Twisted Pair — Data Out

Noise In — Nothing Out

Figure 2-35

Data are sent complementary, noise is not. The triangular device is an operational amplifier (op amp). The negative (-) input inverts the signal in the op amp. The signals are then added to create a greater amplitude output signal. The negative signal on noise is inverted which results in a signal of opposite polarity. Opposite signals will cancel in the op amp.

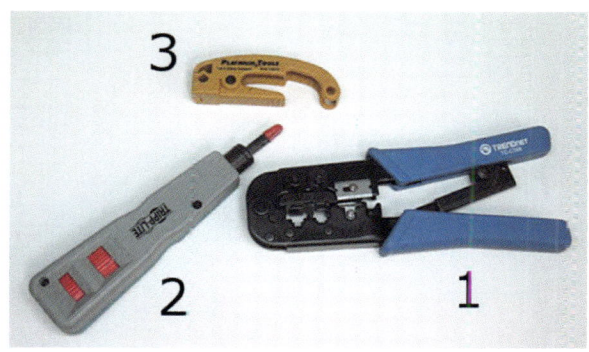

Figure 2-36

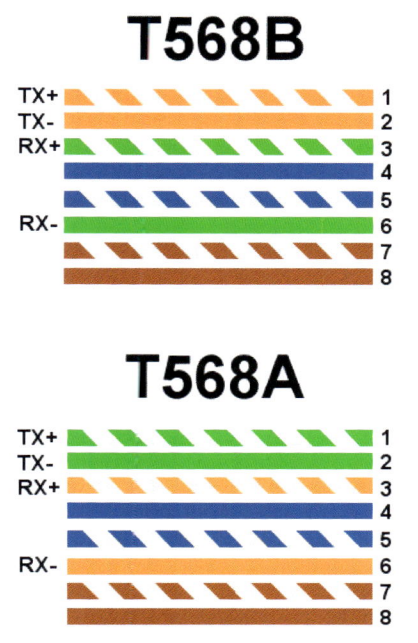

Figure 2-37

Cat5e cables have a bandwidth of 100MHz. Cat6 cables have a bandwidth of 200MHz and are less susceptible to electrical noise. Think of bandwidth as a road. Doubling the size of the road will double the amount of traffic or, in our case, data. The improvements Cat6 allow also increase reliability. Cat6 is also backwards compatible, i.e., it will work in a Cat5e system.

Cat5e cable and a Cat6 connectors and plugs cable are made the same way and the wireing order is the same. Only one tool is needed to make a jack, a Cat5/Cat6 crimping tool (item 1 in **Figure 2-36**). A Cat5e/Cat6 *cable stripper* (item 3 of **Figure 2-36**) is nice to have but not absolutely necessary. To make a RJ-45 plug, a *punch down tool* is needed (item 2 in **Figure 2-36**).

There are two wiring sequences standard to the RJ-45 which were established by the American National Standards Institute/Telecommunications Industry/Electronic Industry Alliance (ANSI/TIA/EIA). The standards are ANSI/TIA/EIA -563-B.1-2001, -B.2-2001, and -B.3-2001. The styles are 568A and 568B wiring hierarchies. Both wiring orders are seen in **Figure 2-37**. Cables can be made in two styles, *straight thru* or *crossover*. A straight thru cable is one where both ends of the cable are attached to plugs (or jacks) using the same standard 568B is on one end and the other end is also a 568B connection. Crossover cables will have a 568B on one end and a 568A on the other.

Start the cable by stripping two or three inches (5 to 7.5 cm) of sheathing from the table. You must not damage the insulation on any of the twisted pairs. When removing the sheathing, be sure to leave the wire pairs in their original twisted condition, or lay. Removing the twist increases their susceptibility to electrical noise. The cable stripper in **Figure 2-38** will cleanly cut the sheath and leave the wires inside untouched. If you use another tool to strip the wire and are unsure if the wire or insulation are damaged, take the stripped end of wires, hold tightly, and pull the nylon cable away from the gripped wires. The nylon will split the sheath and expose fresh, undamaged, wire.

Figure 2-38

A dedicated cable stripper for Cat5e and Cat 6 cables. Simply open the clamp, attach it around the cable, and rotate the device around the cable two or three times.

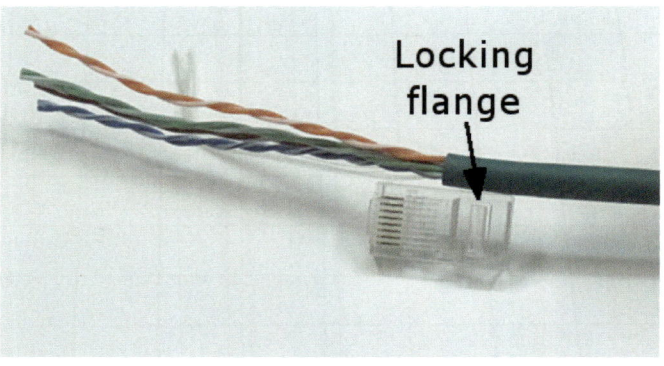

Figure 2-39

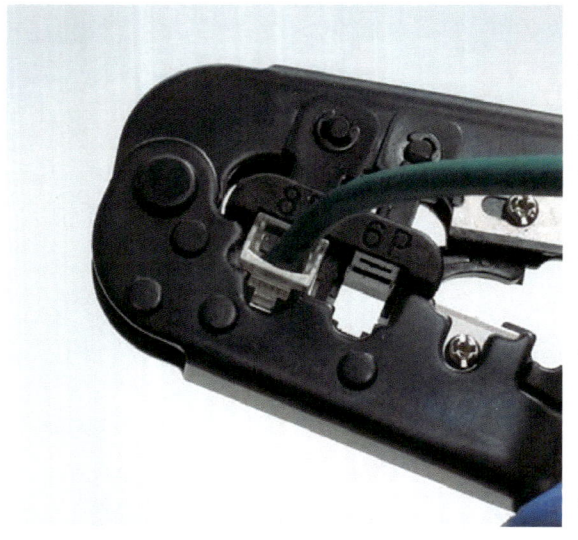

Now that the cable is stripped, place it next to the RJ-45 connector to determine where the wires must be cut. There is a small plastic locking flange, marked in **Figure 2-39**. This clip is pressed onto the outer sheath to anchor the cable. This prevents tension on the wire ends connected to the pins. The sheath must go under the clip. The wires must be cut to fit to the end of the connector and allow the sheath to be under the flange. When you arrange the wires in the correct sequence, try no to untwist them. This is difficult to do on the plug but is much easier on the jack. When you have the wires in the proper sequence, carefully place the plug in the crimper (it will only fit one way). Make sure the wires are seated into the plug as far as possible. Simply close the crimpers and they

Figure 2-40

will lock the sheath and force the copper teeth on the jack to make contact with the wires inserted into them. A properly crimped plug is shown in **Figure 2-41**. **Figure 2-42** is a poor crimp because the sheath is not under the locking tab.

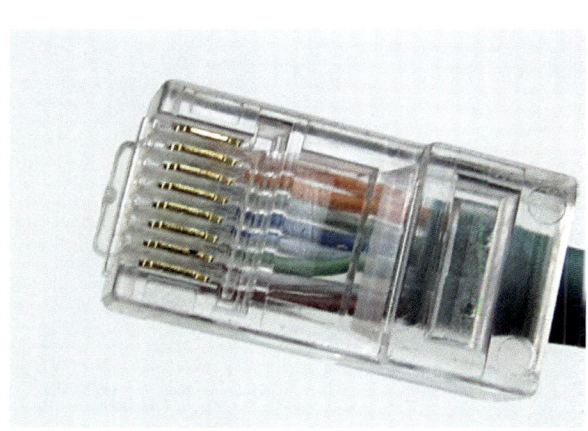

Figure 2-41
A proper crimp

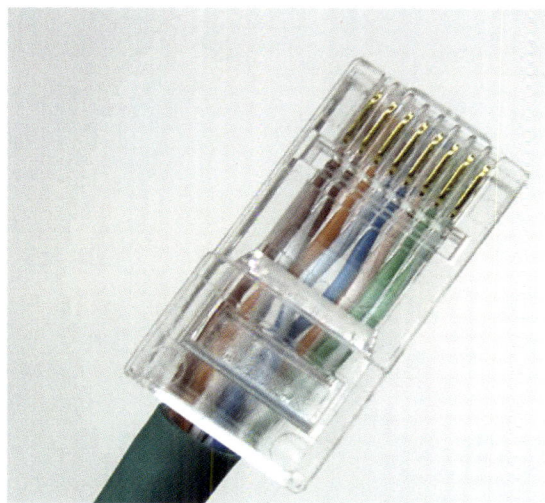

Figure 2-42
A poor crimp, the sleeve is not under the locking flange.

Figure 2-43

This is a connector featuring a stress relief boot. It further reduces tension on the wires in the jack.

The cable for a jack is prepared the same as for a plug. You must have a punch down tool to make a jack. Again, the punch down tool is item 2 in **Figure 2-36**. Familiarize yourself with the tool before you use it. The punch down has two functions, to force the wires into the jacks connectors, and to cut the excess wire. One side of the tool tip has a cutting edge to trim the excess wire. This cutting side, seen in **Figure 2-44**, should be on the outside of the jack when the tools is used.

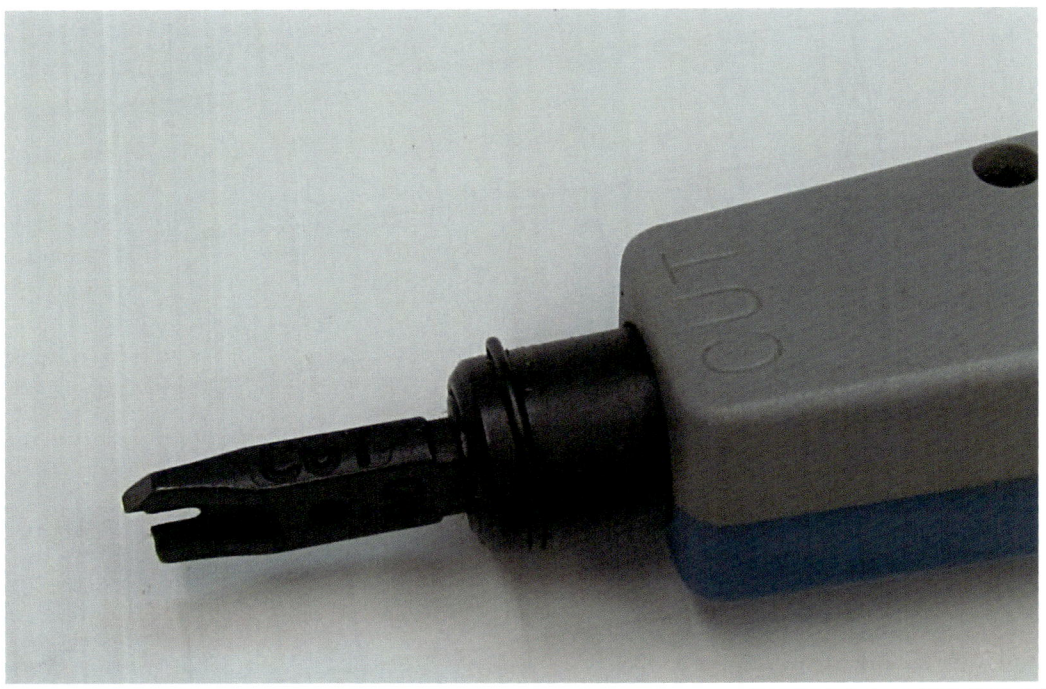

Figure 2-44

Place the cable in the center of the module. Keep the leads twisted and connect the wires to the correct slot using the shortest length of wire necessary. The wire should remain twisted in their original configuration to maximize noise immunity. When the wire is in the correct slot, place the punch down tool above it (with cutting edge out) and press down on the tool. The tool has a plunger that will actuate when sufficient downward pressure has been applied. The spring actuation forces the wire into the pins and cuts the wire at the same time.

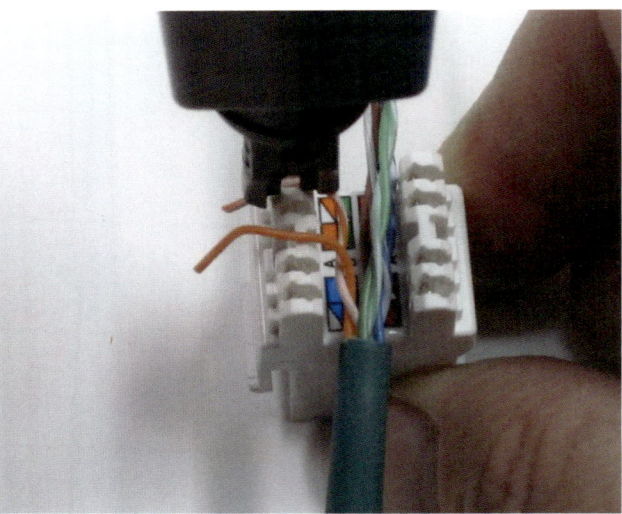

Figure 2-45

A completed jack is shown here. Note that the wires all terminate flush with the jack and that the wires are at the bottom of each slot. A wire that is not at the base of its slot is not making a connection with the pin below it.

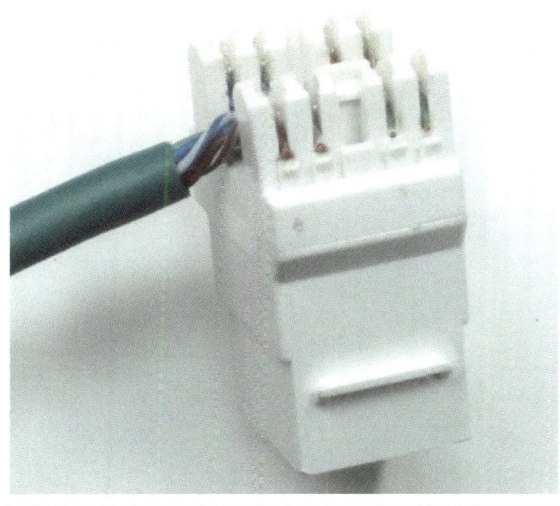

Figure 2-46

Chapter 3

Irons, Solders, and fluxes

Soldering is the process of joining two or more electronic components together to form a functional circuit. This chapter discusses the principle tools of soldering, the soldering iron, solders, fluxes, and cleaners, and begins to introduce some of the concepts behind high reliability soldering.

What is Soldering?

Soldering is the process of joining two materials by heating a filler material without melting the metals of the components. A filler material, the solder, has a melting point of 840°F (449°C) or less. This is usually below the melting temperature of either of the two metals that are being joined. It is this temperature that distinguishes soldering from brazing and welding. In brazing, the metals are joined by heating filler and base metals to over 840°F (449°C) but below the metals melting point. It is advantageous to braze when the two base metals are of different types and not too large. This is because brazing usually heats all of the material to be joined. Welding, on the other hand, actually melts the base metal to cause fusion of the two pieces (a filler material may be used). Welding is most useful when joining large assemblies and when heat can be applied to relatively small areas.

The Tools

The Soldering Iron

A soldering iron is a hand-tool, usually electrically powered, consisting of a heating element, heat reservoir, and tip. Tip types and sizes vary according to use. This is also true of the heating element – different wattages provide varying degrees of heating capacity. **Figure 3-1** shows an iron that is acceptable for most basic soldering tasks. The tips are interchangeable to

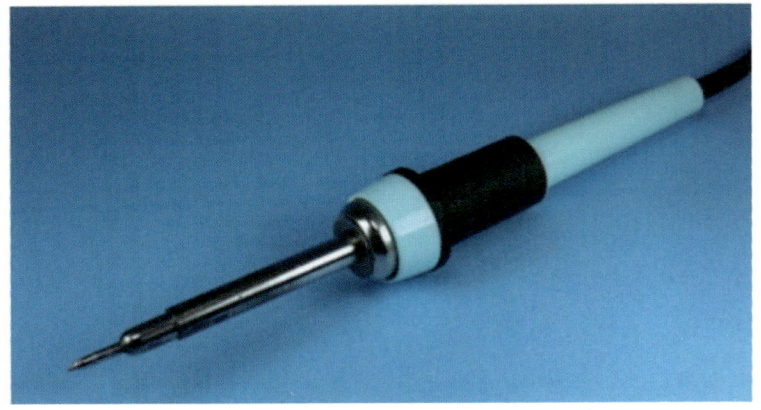

Figure 3-1

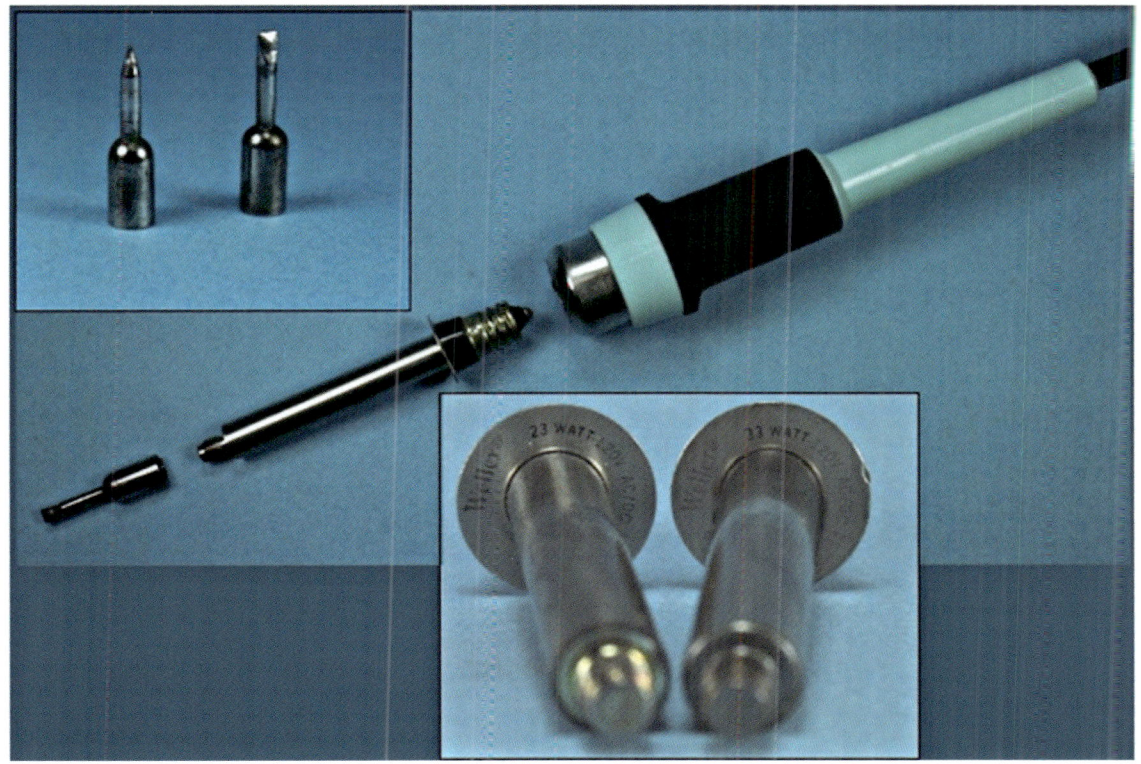

Figure 3-2
A soldering iron that has replaceable tips and heating elements can be used in a variety of applications.

allow use over a broad spectrum of component types. Additionally, the heating element and heat reservoir are also interchangeable. This flexibility lets the technician adapt the iron to the size of the work being done.

Figure 3-2 shows the iron disassembled into its functional elements; tips (top left inset), heating element and heat reservoir (combined in one unit, bottom center inset), and the handle.

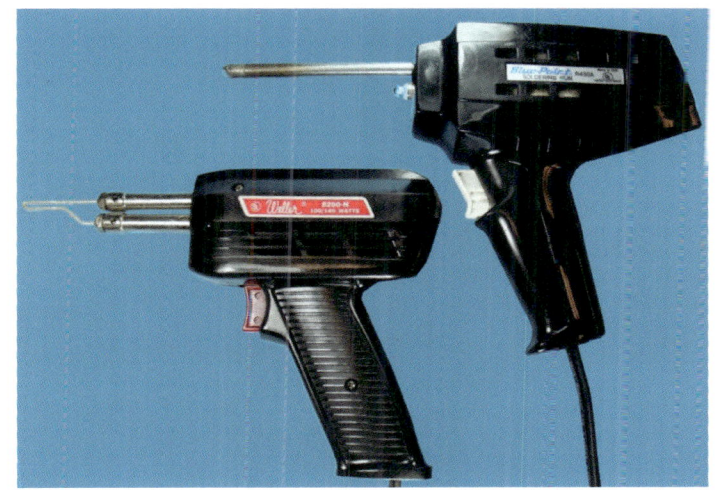

Figure 3-3
Excellent tools for their intended purpose but not to be used in high-reliability soldering.

The two tips in **Figure 3-2** can be used for the great majority of electronics work/rework. The conical tip (left side of top inset) is excellent for soldering and desoldering through-hole integrated circuits (IC) in addition to many surface mount devices (SMD). ICs and SMDs will be discussed later. The chisel tip to the right in the upper inset works well for larger tasks such as soldering 24AWG or larger wires onto circuit cards, connector terminals, tinning, etc.

Two heating element/reservoirs are shown in the lower image of **Figure 3-2**. The heating is accomplished by running a current through a resistive wire made of nickel and chromium that is wrapped inside the heating element. This alloy, often called nichrome, has a very high melting point (around 2550°F - 1400°C) and does not oxidize easily; the two principle reasons for its use. Nichrome wire tends to be expensive because of nickels high cost. Various wattages can be attained at the factory by varying the number of turns of nichrome wound around a core. The 23 watt element can reach a temperature of approximately 700°F (371°C). This is an excellent temperature for those new to soldering. The 33 watt element will reach about 800°F (427°C) because it has more windings of nickel/chromium wire than the 23 watt element. However, because of the higher temperature reached by a 33W element, board damage can occur quickly when used by an inexperienced technician.

Figure 3-4

Butane powered soldering iron. A good tool for fieldwork.

The winding of resistive material onto a central core is the most common way of making a soldering tool. Other methods used to generate (and, to some degree regulate) temperature include resistance soldering, use of transformers or switching power supplies, butane soldering irons, and soldering guns. Soldering guns should never be used in circuit repair. It is a task they are not designed for and the hapless person who does try to use one will probably cause so much damage that the entire board must be replaced.

Solder

Solder acts as the filler material in the joining process. The most common type in the United States is composed of a mixture of tin and lead. Any roll of tin/lead solder will have the proportion of the two metals stamped on its side. A roll stamped 50/50 would consist of 50% tin to 50% lead. The tin value is always listed first.

There is a reason for the use of a tin (symbol Sn) and lead (symbol Pb) amalgam. Pure tin has a melting temperature of 449°F (232°C). Lead melts at 621°F (327°C). However, when both are mixed. the melting point of the resulting alloy is less than the melt point of either of the constituents.

This can be seen in the following chart (**Figure 3-5**). For example, an 80/20 ratio would have a melting point of approximately 550°F (288°C). A 20/80 ratio would melt at slightly over 400°F (204°C). Metals mixtures can have one other quality – the mixture can be in a **Plastic State**. When in the plastic state, the solder is not completely molten (in its *liquidus* state) nor is it completely solid (the *solidus* state). This is an important point to remember. When working with solder, make sure it is completely molten when joining the two items and make sure the work is not allowed to move during the cool down period. If either of these points is ignored, a deformed connection will be made and a high reliability joint is out of the question.

You may have noticed the point where the plastic areas meet. This point, 361°F (183°C) for Sn/Pb solder, is called the **Eutectic Point**. At 361°F (183°C), the solder goes directly from a solid to a

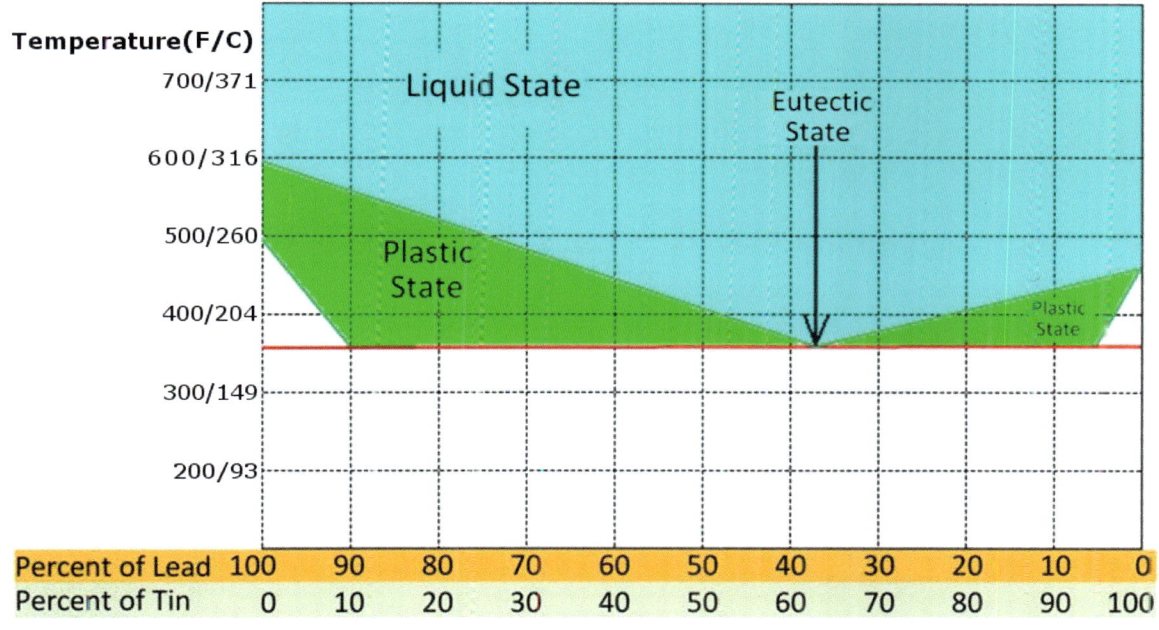

Figure 3-5

This chart shows the relationship between various tin/lead solder ratios and their melting points. Notice that plastic ranges also vary.

liquid – there is no plastic range. For Sn/Pb solder to be eutectic, it must have a 63/37 ratio. Eutectic solder is most useful where a chance of component movement during the soldering process exists. It also has benefits when heat sensitive components must be installed. Because of its relatively low melting point, it melts quickly subjecting components to heat for the shortest possible time.

There is one final point to be made about eutectic solders. Many alloys exhibit a eutectic state. Don't be surprised to find other solder alloys listed as eutectic. Remember, eutectic simply describes a material that goes from solid to liquid (or the other way around) without having a plastic state.

There has been much interest in the past few years in removing lead from solder. This is because the European Union enacted the **Directive on the Restriction of the Use of Certain Hazardous Substances in Electrical and Electronic Equipment 2002/95/EC**, more commonly called RoHS (Restriction of Hazardous Substances). This policy became effective July 2006 with the purpose of removing six hazardous materials from electronic equipment. Lead was at the head of the list with mercury and cadmium following.

Because of the removal of lead, new solder compositions have been introduced. These new solder alloys tend to have a higher melting point. They also tend to be more difficult to work with because of 1) the higher temperatures involved and 2) they do not flow as well as tin/lead solders. Even experienced technicians can have trouble with these new alloys!

Solder composition is also adjusted for the materials to be soldered. If low temperature soldering is required, bismuth (Bi) will be added. Silver (Ag) is added in joints requiring high strength. Silver is quite expensive so substitutes are being sought. Copper (Cu) is one of these, SnCu alloys are an example. Copper can leach out of the connection during soldering so the SnCu mix was developed reduce this problem. Antimony (Sb) is another metal found in solders. It is principally used in lead free solders such as SnSb. Its benefits are strictly environmental as it is subject to poor joint connections when

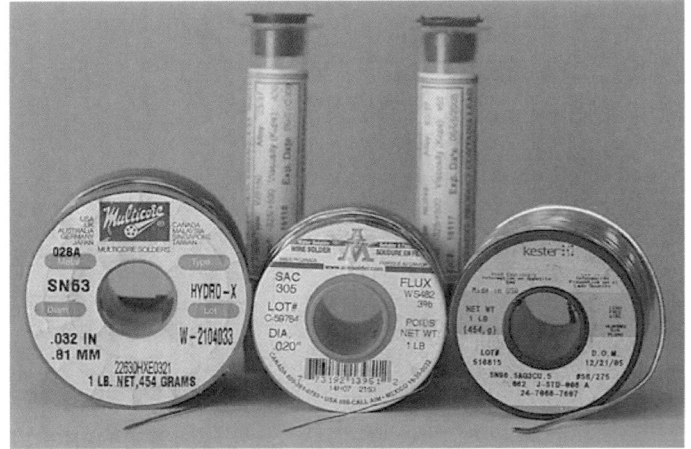

Figure 3-6
Three types of roll solder. Paste solder is seen in the back.

used as an alloy in solders. **Figure 3-6** shows some currently available alloys.

Solder comes in two common forms; rolls (which usually weigh a pound) and as pastes dispensed in syringe like tubes. The pastes is both solder and flux (flux is discussed below) evenly mixed in a jar or syringe. The roll solder is a long section of solder coiled onto a spool. The solder size can be between 0.010" to 0.050" (0.25 mm to 1.27 mm).

Paste solders have a shelf life ranging from 3 months to a year. The tin/lead mixes are useable for 3 to 6 months; lead-free solders are good for one year. To achieve this shelf life, the pastes must be refrigerated. Additionally, the paste must remain sealed because it is *hygroscopic*, it absorbs moisture. Leaving a solder paste container open will cause the solder to oxidize. Finally, solder paste is *thixotropic*. This means its viscosity decreases over time. It should be mixed before use, if possible. The problems of changing viscosity are of critical importance in automated manufacturing. If the viscosity increases, the amount of solder paste needed at one point cannot be achieved because it flows away.

Flux

The majority of solders also contain a flux. The purpose of flux is to remove oxides from the surfaces to be connected and to prevent new oxides from forming during the soldering process. Technically, flux is a *reducing agent*. It reduces the

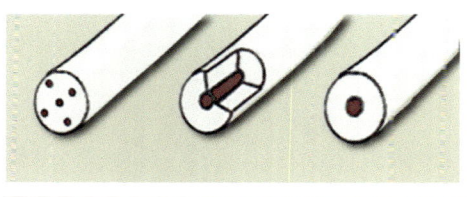

Figure 3-7
Flux distribution in solder.

oxidized metal to its original state. It also aids in *wetting* – the process by which solder flows onto and into the materials to be connected. Solder does flow into the base material it is used on. It is in this way that an *intermetallic bond* is formed leading to the excellent mechanical and electrical characteristics of a well soldered connection. The intermetallic bond will be discussed in greater depth later. If you are using solder from a roll, flux is found in the center of the solder or spaced evenly about the center as shown in **Figure 3-7**. Additional flux can, and more often than not, should be added. It can be dispensed with a bottle and small brush, pen style devices (much like markers), plastic bottles with needles, and with a brush (or finger) from a flux paste jar.

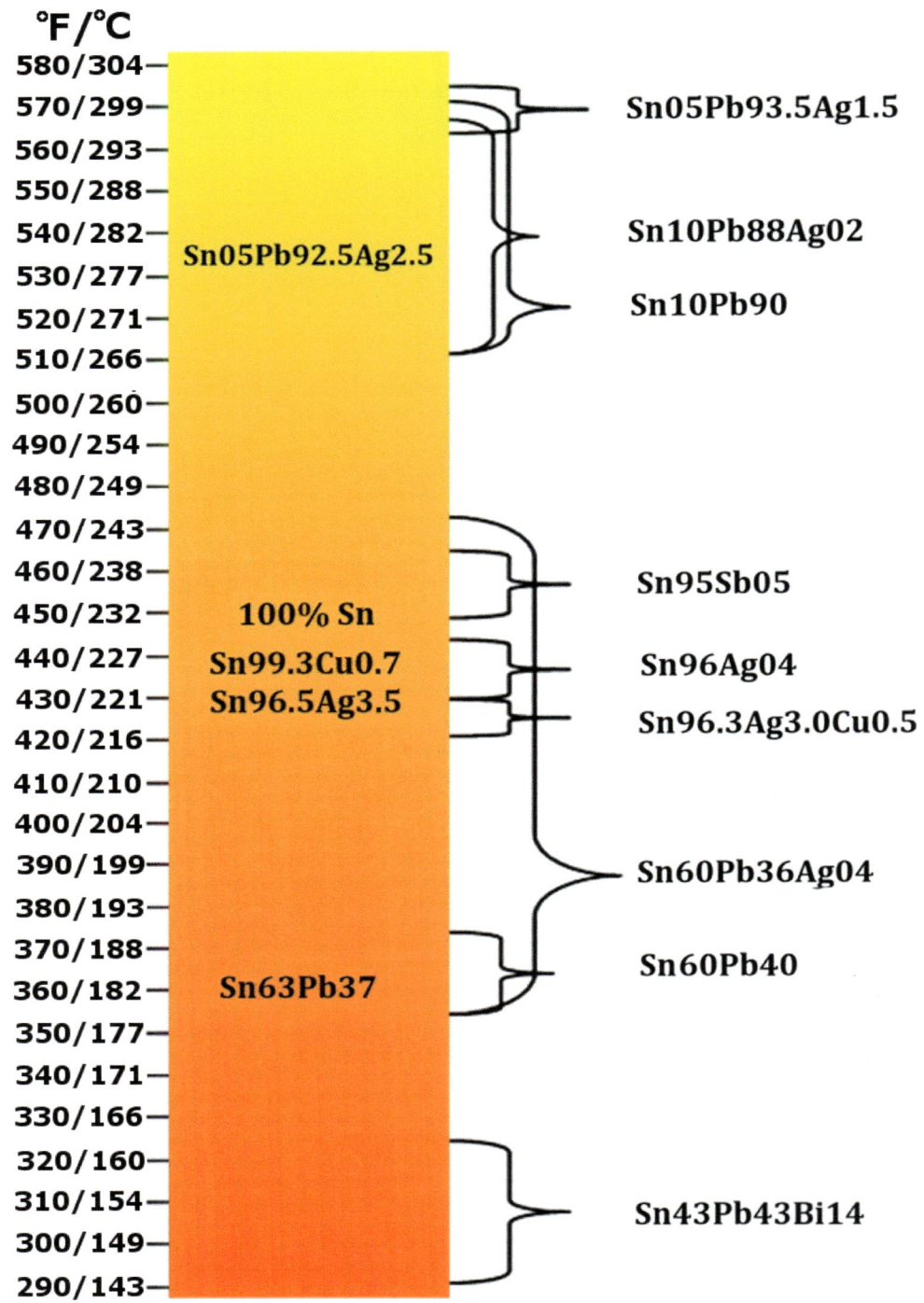

Figure 3-8
Solder alloys inside the colored temperature band are eutectic – those to the right are shown with the plastic ranges bracketed. All temperatures are in degrees Fahrenheit.

Solder fluxes fall into two categories: rosin core and water-soluble. Rosin fluxes are made from extracts of pine sap. The resulting material is an organic acid. Because of this, rosin core fluxes may be referred to as organic acid (OA) fluxes.

Flux must have an activation temperature lower than the melting point of the solder. After all, its purpose is to remove oxides before the solder melts and seals the connection.

Solder comes in a variety of sizes. The majority of manual soldering can be accomplished using 0.025" to 0.050" (0.254 mm to 1.27mm) diameter solders (0.031" (approximately 0.8mm) is a good "all-around" solder size). Because of the size variation in solder, there has to be some way of determining the volume of flux contained in the solder as a percentage. Manufacturers have established three (3) standard values; 1.1%, 2.2%, and 3.3%. So, a 0.031" (0.8 mm) wire with 1.1% flux has that amount of flux by weight of the solder. No matter what the solder size, the amount of flux is known based on this percentage. It is important to know the flux percentage. The more contaminated the work piece is, the higher percentage of flux required for proper cleaning. Acid core flux is also be available. _**Never**_ use this on electronics! It is used in plumbing or metal fitting and will destroy the printed circuit board and components.

Quite often, other compounds, such as isopropyl alcohol and/or halides, are added to aid the solder process. The use of halides is questionable. It does make the flux more active but it also has its own negative charge so it becomes an ionic contaminant. This can cause leakage current. The types of materials added determine the class of flux you are dealing with:

RO – rosin only. This is very gentle flux and is easy to clean as it leaves few solids behind. It can only be used where very clean contact surfaces already exist as it cannot remove even moderate oxides. This flux contains few added acid activators.

RMA – rosin mildly activated. RMA is slightly more active than RO and its residue must be removed after use. Cleaning, however, is relatively easy. This is good general use flux.

RA – rosin activated. Very active and used where oxidation is extreme. Unfortunately, it also leaves considerable residue which has to be cleaned.

(IN) - Inorganic and synthetically activated fluxes (RE) also exist. IN solders are used in industrial applications to solder brass, copper, and stainless steel. They have recently been introduced into electronic soldering.

The water-soluble (WS) fluxes are IN types and are very active. Water-soluble fluxes, despite their name, are not water soluble until activated in the soldering process. Water-soluble flux left on the work after soldering remains active. It will continue to remove oxidants as well as the base material with which it is in contact. This can lead to degradation of the contact. Any water-soluble flux that remains after soldering will have to be removed with the appropriate solvent (filtered water, and/or de-ionized water). The residue left by WS fluxes can also change the measured resistance of a point. This can cause a board to appear bad during testing when it may be fully functional. The use of filtered or de-ionized water is a highly desirable feature of WS fluxes. No volatile organic compounds (VOCs) are generated when water is used as an ingredient of flux and as a cleaner. Alcohol, which is an ingredient in many fluxes and cleaners, is a VOC. The absence of VOCs makes WS fluxes environmentally friendly. WS flux does have the disadvantage of needing higher temperatures to activate and burn off. Remember, water boils at 212° (100°C) while alcohol needs much lower temperatures, between 60° and 70°F (15.5°C to 21°C).

Fluxes are now being sold that have a "no-clean" (NC) characteristic. These fluxes tend to be mild so more time is needed for them to activate and allow proper wetting. These fluxes do leave a residue behind which is supposed to be electrically and chemically inert. However, under some conditions, like automated resistance testing or stencil application, cleaning is still required. Also, if the board is to receive a conformal coating, the residue must be removed. NC fluxes may leave a white powdery residue or what looks like an oily liquid that does not dry. The liquid residue can collect dust and debris. Cleaning is done with solvents or an alkaline water solution (aqueous saponifier). Some NC fluxes cannot be cleaned at all.

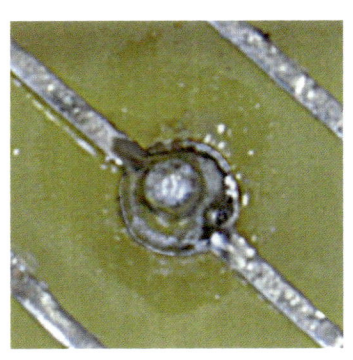

Figure 3-9
A No-Clean flux was used in this solder connection.

A quick point on no-clean fluxes. The connection in **Figure 3-9** was done with lead free solder containing NC flux. This is an example that NC does not mean no residue. In some cases, the residue will be white and powdery as opposed to this wet almost oily look. Recall that the only requirements for NC fluxes were that they be non-conductive and non-destructive (i.e. causing board or pad/trace degradation).

All fluxes work the same way. When you place the flux (an acid) onto an oxidized base material, salt and water are the result. Both of these resultants must be removed to prevent possible degradation of the board. Leaving the fluxes' solids (the solids are the part of the flux that is active) behind can cause changes in board resistance and the formation of undesirable circuit paths.

The type flux that can or should be used is dependent on two elements: 1) how active must the flux be to perform the soldering task and 2) what level of cleaning, if any, is the product to undergo after soldering. The first point is obvious. Since lead free solders don't work as well in wetting electrical contact areas as those containing lead, a more active flux must be used. The second item is based on the working environment of the product and the reliability required of that product. Humid environments require that fluxes be removed completely. Flux can also form branches, called dendrites, to other contacts and lead to short circuits. It can also change surface insulation resistance (SIR) which, as mentioned above, leads to inaccurate test results or worse, a board that fails.

The level of cleaning is also dependent on the reliability you want out of your equipment or that is required of that equipment. Three classes of electronics are defined by IPC/EIA J-STD-001D:

CLASS 1 – The product must function.

CLASS 2 – Service of the item should be uninterrupted but, in the event of failure, the function of the item is not critical.

CLASS 3 – The equipment must work at all times in the environment for which it is designed. Medical equipment and aviation electronics (avionics) would fall into this group.

The line between Class 1 and 2 is somewhat blurred. Here the manufacturer has leeway in the removal of flux residue and the quality of the solder joint. Class 3 is unambiguous. The highest standard of construction and cleanliness must be maintained. IPC/EIA publishes another standard (J-STD-004B, Requirements for Soldering Fluxes) whose sole purpose is to describe the requirements of soldering fluxes and their suitability in various production methods.

Throughout this text, the Class 3 standards are used whenever applicable. If you can achieve the Class 3 standard, you have already mastered the Level 1 and Level 2 standard.

Classifying Types of Flux

IPC developed standard J-STD-004B in part to make quick comparisons between various types of flux easier. The classification method uses three elements: 1) is the flux RO, RMA, IN, RA, as described on pages 62-63, 2) how active is the flux and, 3) Yare halides present. You are already familiar the first classification group. The next classification group determines if the flux has low activity (L), medium activity (M), or high activity (H). The amount of activity is determined by the percentage of halides in the flux. L = < 0.5% halide *or* low activity, M = 0 to 2% halides *or* mildly active, H = 0 to 2% halides *or* highly active. An indicator is also included to let the potential user know if halides are present. Zero indicates no halides while one indicates the presence of halides. A ROL0 would indicate a rosin solder that has low activity and contains no halides.

The How and Why of Soldering

As was mentioned earlier, soldering is the process of joining two metals to make an electrically invisible but mechanically strong connection. When a good connection is made between the work pieces, an intermetallic bond is created. The new metal is part copper, part tin and perhaps leads, copper, or silver (depending on the type of solder used). An illustration of this is shown in the **Figure 3-10**. The bottom area is a fiberglass board which is typical of most current electronic construction. The material above represents the copper pad which must be tinned using solder. Tinning is the process of applying a thin coat of solder to a contact point. It makes soldering to a connection easier because the tinning prevents oxidation and helps limit contamination of the leads, pads, etc.

For proper tinning to occur, cleaning of the part with a solvent (alcohol or other appropriate cleaner) must be done. A small amount of flux is then applied to the part to be tinned, the flux is heated by the soldering iron to activate it, and then solder is applied. If the part was not cleaned, the connection may look like the left pad of **Figure 3-9**. The solder has essentially formed a ball which sits on the copper pad. Just a small amount of mechanical strain would cause this connection to fail. The image on the top right shows a proper connection. Note that the copper and solder have blended together and that the copper has been completely covered. This blending makes the joint extremely strong.

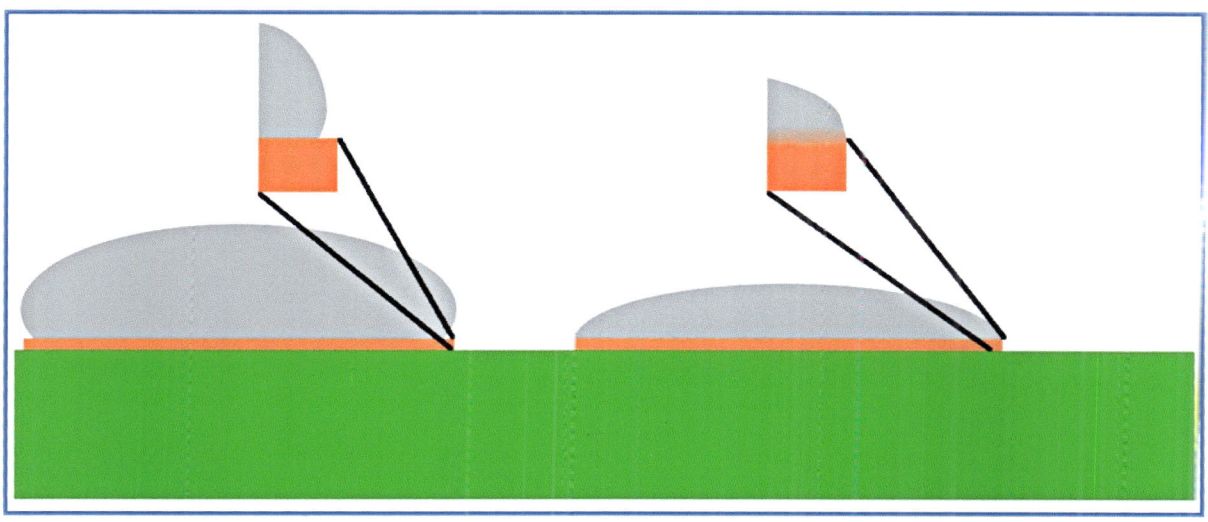

Figure 3-10

A poor solder connection is shown to the left. An intermetallic has not been created. The preferred connection in which solder and base metal have "blended" is seen on the right.

Mass, Time, and Pressure

Whenever you perform a soldering task, you must be aware of three things: the size of the work piece, the pressure you apply to the work piece, and the amount of time your iron contacts the work piece. Any other points are sub-categories of these elements.

Mass - Work Piece Size

The size of the item you are soldering will affect dwell time (the amount of time needed to achieve flux activation, solder melt, and a good connection). The bigger the mass, the more power it will take to get it hot enough for solder melt to occur. The power of the iron (wattage) and its ability to recover from heat loss (the size of the heat reservoir) become important here. If the heat reservoir is too small, heating will take much longer than preferred. This can lead to board and component damage. If the soldering iron's tip is too small, similar damage can occur. **Figure 3-11** shows a turret terminal and bifurcated terminal. The turret is considerably larger and requires the larger tip

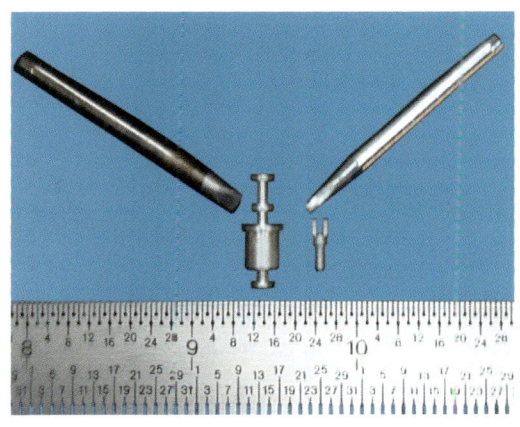

Figure 3-11

Different masses (turret terminal to left, bifurcated on the right) will require different size tips.

for quality soldering. The key to tip selection is the amount of thermal linkage between the work piece and the iron. Thermal linkage is the area of contact between work piece and iron. A larger tip offers more contact area.

Using a tip that is too big or an iron with too much wattage can be equally disastrous. In the

Figure 3-12

Only the third and fourth tip in this example would be suitable for use on the board.

center of **Figure 3-12** you can see five small pads circled in red. The two soldering iron tips to the left are definitely the incorrect size. They are too wide. These tips could lift the pads off the board or cause a problem called *measles*. Most circuit cards are made of fiberglass and epoxy. If the circuit board were overheated, the epoxy/fiberglass would separate (because the epoxy has over cured) and small white spots will form – these spots are measles. This is considered a serious defect in Class 3 equipment. The board would require repair or have to be replaced.

Time

After the correct tip has been selected, you have to concentrate on the amount of time you are in contact with the work piece. Too long a time and you have measles, too short and you have a *cold solder joint.* A cold solder joint will appear grainy and dull. With the right tip, it should take about two to four seconds to finish the work. This provides enough time for the flux to activate and the solder to wet the component.

Pressure

The amount of time you can spend in contact with a connection, as you have noticed, is variable. Individual techniques have great impact on the time variable. However, the amount of pressure that a technician applies to the work piece is not variable. The copper on a circuit board is held on the fiberglass substrate with an adhesive. Heating during the soldering process decreases the adhesives bond strength by up to 80%. Applying pressure in addition to the heat can easily exceed the glues' diminished strength. You then end up with a lifted pad or trace. The amount of pressure applied with the iron during soldering should only be enough to keep the iron in place. Another way to get the feel for this is to imagine the iron being a pencil with a very sharp tip. If you apply too much pressure the tip of the pencil, it breaks. This is the maximum amount of pressure you should apply with the tip of the iron.

Some Basic Tools

A set of tools that would work for the majority of through-hole soldering (the component is mounted through a hole in the circuit board) and much surface mount work are seen in **Figure 3-13**. This equipment doesn't have to be costly. In many instances, the greatest costs are the expendable items such as solder. By adding just a few items, you can expand into board repair (**Figure 3-14**).

Tool List

Figure 3-12 has all of the tools listed below.

A	Vise and stand
B	Circuit board vise
C	Common and Phillips screwdrivers
D	Pliers
E	Safety glasses
F	Lint free wipes
G	Paste flux
H	Flux removers
I	Fluxes
J	Assorted types of solder (eutectic, lead free)
K	Soldering iron
L	Soldering iron stand, tray, and sponge
M	Conical and chisel tips for soldering iron
N	Solder wick
O	Tip tinner
P	Solder extractor
Q	Cleaning brushes
R	Manual wire strippers
S	Steel ruler
T	Dental pick
U	Abrasive stick (white eraser)
V	Orangewood sticks
W	Lead forming tool
X	Hobby knife
Y	Tweezers
Z	Round nose pliers
AA	Duckbill pliers
BB	Coarse and fine wire cutters
CC	Needle nose pliers

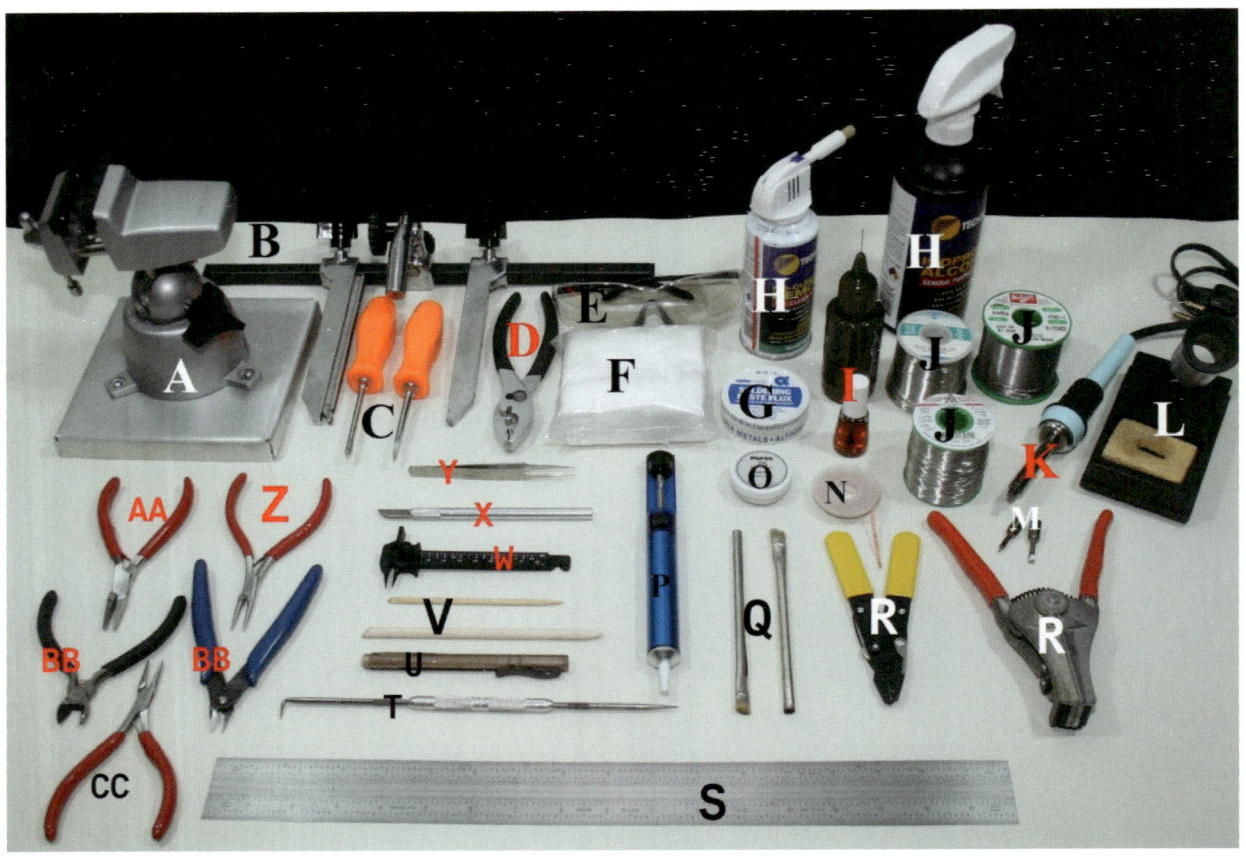

Figure 3-13

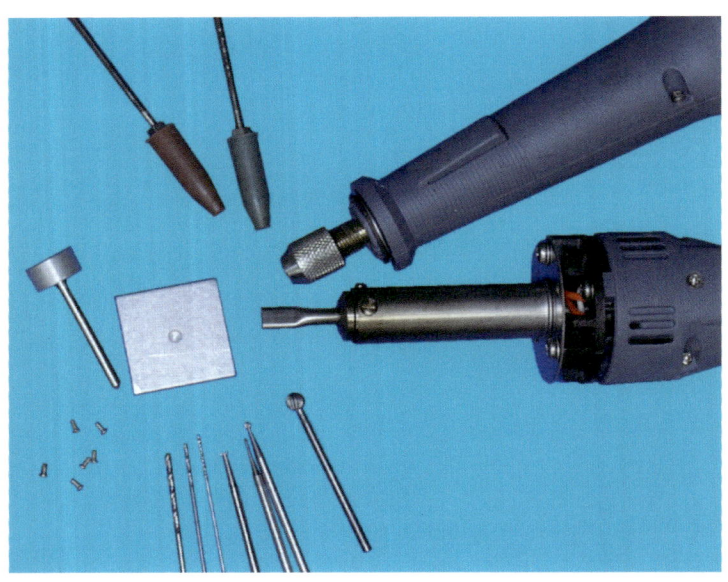

Figure 3-14
A small selection of tools used for more advanced work on printed circuit cards.

If you are working on electrostatic sensitive devices (ESD), an ESD mat and ESD wrist or ankle strap will be needed.

Fume extraction is necessary in enclosed areas so a fume extraction device should be used. And, if ESD is an issue, the device should be approved for ESD applications.

Later chapters will cover the repair of circuit boards. This will require a few additional tools above the basic set. A rotary machine like a Dremel® or

Microchine© will be needed to be obtained. A hot air system is a plus. **Figure 3-13** shows a fairly representative set of tools needed for more advanced repairs. Shown in this image are; brown and green abrasive bullets, a Microchine©, a hot air jet, several sizes of ball mills, a small rotary saw blade, a selection of drill bits, some sample eyelets for board repairs, and a swaging tool and base.

Chapter 4

Wire Connections and Connectors

This chapter begins the actual practice of soldering. We will start with tinning wire and progress through several types of wire connections. Each connection will have different properties which will enhance your understanding of the effects of thermal mass and dwell time during the soldering process. You will start *reading* the work and adjust your technique to achieve high reliability connections.

Tinning

Tinning is the process of covering an electrical contact with the proper amount solder. It has two purposes: 1) it improves solderability and 2) it prevents oxidation of the contact. Any wire that will be attached to a connector or spliced to another wire should be tinned. Tinning must be done before the wire is connected or spliced.

Achieving proper tinning is something that will take practice. Many mistakes can be made that will require rework or, worst case, replacement of the work piece. **Figure 4-1** shows what occurs when your soldering iron is too cold, too small, or making poor contact with the work piece. Additionally, the wire was not cleaned sufficiently. The wire should be shiny *before* it is soldered. If it has the bluish hue, as exhibited by some sections of this wire, it has oxidized. Such heavy oxidation is the result of being exposed atmosphere for too long or being handled too much by the technician. The oils on fingers are contaminants. It is a good idea to clean and solder such connections as soon as the wires are stripped. Avoid touching the bare conductor and if it must be touched, clean it again.

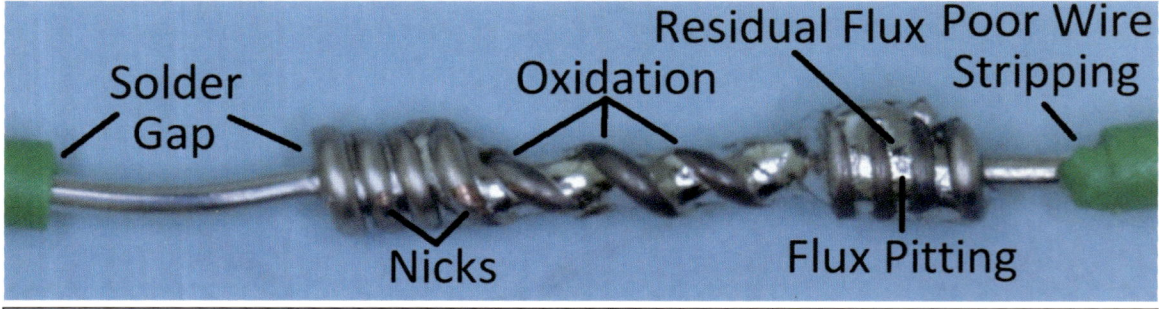

Figure 4-1
Almost every error that can be made in soldering wires is exhibited in this image of a section of a Western Union splice.

The unburned flux in **Figure 4-1** is a definite indicator of cold soldering. Flux activates at a far lower temperature than solder and it should have burned off completely. The unburned flux can cause flux pitting. Flux pitting occurs when the solder has cooled into its plastic state and unburned flux remains. This is similar to oil on water; the oil is light but will displace some of the water. In our example, the flux is light (relative to the solder) but has displaced some of the solder. When that part with the residual solder is cleaned, a low spot is left on the solder joint. The solder spike on the right side tells us that the operator pulled the iron from the work incorrectly. We also see poor wire stripping, incorrect solder gaps, and nicked wires.

Figure 4-2
The square cutout in the sponge is used to wipe excess or old solder from the iron.

Tinning a wire is a good starting point to acquire the skills necessary for high reliability solder. Obviously, you must start with a clean and undamaged section of wire. ***Always*** clean the part to be soldered with the appropriate cleaner before and after soldering. Pure isopropyl alcohol is a good general purpose cleaner. Denatured alcohol is also good, but be aware that the denaturing agents can leave a residue: a "hard" denatured alcohol, one with a large numbering of denaturing materials, should be avoided. Back to cleaning; if you touch it, clean it again, if you drop it, clean it again. The first rule of good soldering is cleanliness. One more important point, *do not tear or pull solder off a roll, cut it off*. Solder has a precise amount of flux in its core. If you tear the solder to separate it from the roll, you change the solder/flux ratio and it may not work as effectively as it should.

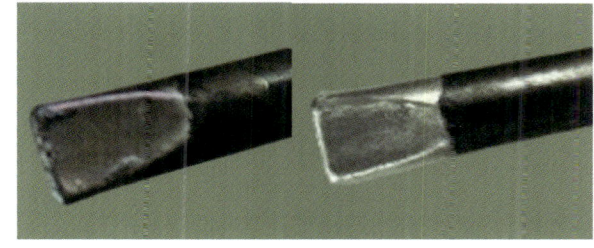

Figure 4-3
The difference between an oxidized tip and a clean tip can be seen here.

The Care and Feeding of the Soldering Iron Tip

A soldering tip can last for years if properly maintained. The tip works in a harsh environment and many routine practices can shorten its life. Use of solders lacking tin, extensive use of solder wicks, and solders with low flux ratios are in this "routine practices "group. Poor practices, i.e., those not associated with routine use, include; wiping the iron on a dry sponge, excessive tip temperatures (800°F, 427°C, and higher), not leaving clean solder on the tip between tasks, and rubbing the tip on the items to be joined. One way to tell if the tip requires maintenance if you melt a small amount of solder on it and the solder forms balls and runs off.

Avoid the items above if possible. If the tip becomes oxidized and shock cleaning doesn't solve the problem, there are several things you can do to try and fix the problem (short of replacing the tip). First, use a large diameter solder (0.050" or 1.27mm, preferably with 2% flux) approximately 5" or 6" (12 to 15cm) long and melt all of it on the tip. Keep doing this until you have only a short piece of solder left. The solder that drips off the tip can will many of the contaminants away. Another method involves using a foam bar specifically made for tip maintenance. The bar has mild abrasives and when rubbed on the tip will remove the oxides. Tip-tinners are also useful. They are a hard paste-like material containing tin and a special flux. Let the iron reach its working temperature and place the tip in the paste for a few seconds. The last option is to use a brass brush and gently wipe the tip. The tip must be at operating temperature when doing this. Bathe the tip in solder as described earlier and then brush both sides gently. Don't use a steel brush as it will damage the plating on the tip.

Preparing to Solder

A good deal of debate circles around the idea of tinning the tip after use. Tinning in this context refers to applying a small amount of solder on the tip and then turning power off. This causes the solder to act as a barrier preventing oxidation. A number of companies claim, and perhaps rightly so, that their tip technology has eliminated the need for tinning. Many end-users disagree. Whoever is right, erring on the side of caution is cheap. A few well-placed drops of solder cost little and may save you the expense and trouble of replacing a tip.

Once a wire is stripped, say ½" (12.5mm) or so, place it in a work vice. Make sure the sponge in the soldering iron stand is moist (not soaking wet) and make sure you have selected a tip appropriate to the mass and size of the item you are working on. A 1/8" (3mm) chisel tip works well for tinning wire. Add a drop of flux to the end of the wire. Capillary action, the ability of one substance to be pulled through another (as water through a sponge), will draw the flux through the wire. At this point, wipe the iron across the opening in the center of the sponge to remove any old solder and some oxidization. The hole in the sponge is there for this purpose – it allows solder to fall into the tray, not on the sponges' surface. Once the old solder has been removed, shock clean both sides of the tip on the sponge. To shock clean the tip, hold the tip of the iron on the sponge for a second or so then flip the iron over and repeat. This removes oxidation and assures that you won't transfer any contaminants to the work piece. Don't leave the tip in contact with a sponge longer than necessary. The sponge can act as a heat sink and cool the tip. If you have used the sponge for some time, it too becomes contaminated. Old solder deposits itself into the pores of the sponge and can reattach to the iron. Replace or clean the sponge regularly. Should the sponge require replacement, use a cellulose sponge, not a manmade one. The cellulose sponge is made from the cellulose of green plants. The manmade sponge is derived from petroleum products, which can melt and release unpleasant and potentially dangerous fumes. When cleaning the sponge, dispose of the solder debris appropriately.

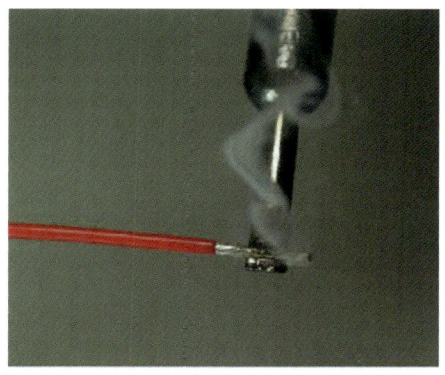

4-4a

Heat the center of the wire.

Tinning Wires

What follows are the general steps for tinning a wire.

Place a small amount flux on the end of the wire and apply solder to the irons' tip, this creates a heat bridge and causes the *wire to h*eat up more effectively. Heat the center of the wire (**Figure 4-4A**). Wait for the flux to activate (indicated by the boiling and then smoking of the flux). Once fully activated, begin adding solder to the center of the wire as in **Figure 4-4B**. How much solder must be fed into the wire is a matter of practice. Begin moving the iron toward the insulation, **Figure 4-4C** (you may still be adding solder at this point) but do not get closer than one insulation gap with the iron. If you do, you might end up with solder that has wicked under the insulation. Wicking cannot occur anywhere the wire must remain flexible. This will vary from organization to organization. Our standard will be that no wicking under the insulation may occur.

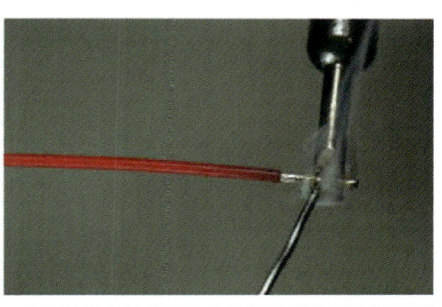

4-4b

Add solder to the heated wire.

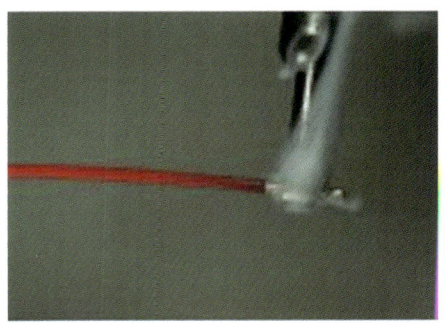

4-4c

Pull the iron towards the insultation but stop about one to insulation gaps from the insulation.

At this point, slide the iron toward the front of and off the wire (**Figure 4-4D**). After cleaning, the tinned wire should look like **Figure 4-4E**. Notice that the strands are still visible. If they disappear, you have used too much solder. In **Figure 4-4F**, the insulation has been pulled back to check for possible wicking under the insulation. You can see the sharp line between the soldered and unsoldered sections of the conductor. The solder flowed only to the spot the iron went. Finally, place a small drop of solder on the iron's tip before storing it.

Recall the three key points of soldering from the previous chapter: mass of the work piece, the time you work on that item, and the pressure you apply. Pressure shouldn't be a problem – if you press too hard, the wire moves. Time is very important here and is closely related to mass. The longer you keep the wire heated and the larger the wire is, the higher the chance of unwanted solder wicking and the greater the chance of damage to the insulation. But, because of the wires mass, more time may be required to heat the wire to proper solder melt temperatures. Tinning wires with gauges of 14, 16, and 18 AWG (1.6mm, 1.3mm, and 1mm) will take practice. Insulation also has a maximum temperature rating that can be easily exceeded with an iron. With potentially increased dwell times, damaged insulation becomes a greater possibility.

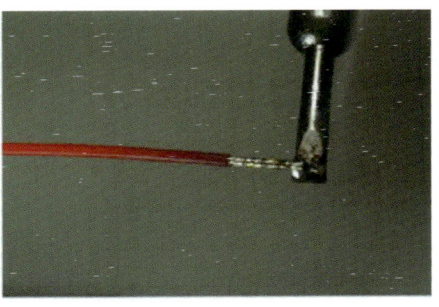

4-4d
Pull the iron off the end of the wire.

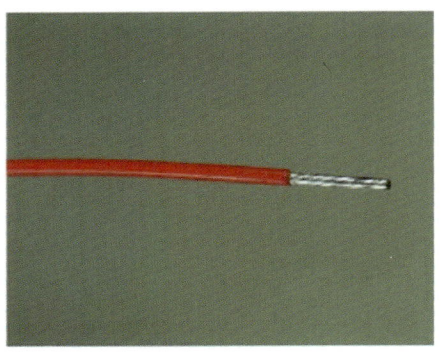

4-4e
You can still see the individual strands in this wire.

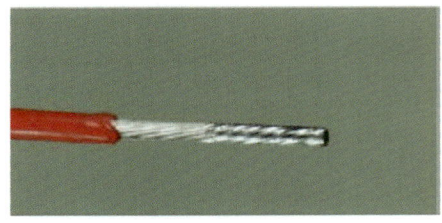

4-4f
There is a very defined point at which the solder has stopped. Preventing solder from wicking under the insulation takes practice.

Another method for tinning wires or electric leads requires a **Solder Pot**. This is a vessel that heats up enough to liquefy a large quantity of solder and maintain it in that liquid state. Some solder pots contain as little as four ounces of solder, others can hold several pounds. The solder pot in **Figure 4-5** is one of the smaller ones.

Usage for all the types is pretty much the same. Once the solder has reached melt temperature, you must skim the **Dross**. Dross is the oxidized and/or contaminated material that floats on the solder surface. Some solder pots will skim the dross with an arm that is moved in a circle around the solder surface. Some solder pots have no skimmer. A common metal tablespoon can be used in this instance to drag and lift the dross out. Once the dross has been skimmed, insert the component to the depth you wish to tin it (do not immerse the component body into the solder) and drag the part sidewise and then lift straight up. A trail of flux and contaminants will be left behind the soldered component. The surface may need to be skimmed of dross again. Component leads should be dipped in the solder pot twice due to the oxidation they accrue while in storage. After extensive use, the solder in the pot should be changed. The contamination eventually becomes too great for any further use.

Figure 4-5
To use a solder pot, dip the lead straight in, drag sideways, and lift straight out. The trail left behind is contaminants that were on the wire. Components that were stored should be dipped twice.

Hooks and Eyelets (Pierced Terminals)

Eyelets are often found on rheostats and potentiometers. Multi-contact relays also sported quite a few eyelets or hooks. Although the hook is becoming rare, it will still be covered. The technique for working on an eyelet is almost identical to the one for working on a hook terminal. An eyelet and hook are shown in **Figure 4-6**. The eyelet, on the device to the left, is part of a Silicone Controlled Rectifier (SCR), a fairly common component.

The drawing of **Figure 4-7** shows how wires are to be attached to eyelets and hooks. Notice the length of the stripped portion of the wire as it wraps around the eyelet and hook – both terminate parallel to the top of the connection. Also, the wire must be in contact with the terminal at all possible points; you always want to maximize the electrical contact area to achieve the best possible connection. This also increases the mechanical strength of the connection once it is soldered. Any contact point between the component and the wire must be soldered. A solder gap between insulation and component must also be present.

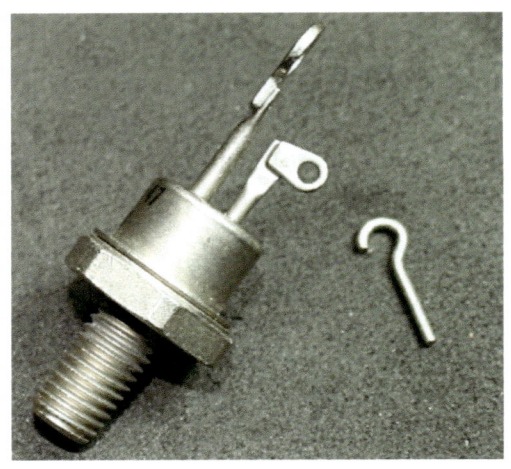

Figure 4-6
Just two examples of the many styles of eyelets and hooks available.

When soldering any component, begin by cleaning and then adding flux. In some cases, extra cleaning must be done. Oxidation of components, especially those that have been stored for some time, can be severe, in some cases great enough to prevent even a fluxed component from wetting during soldering. In such cases, an *abrasive* stick might be used. Abrasive sticks are designed to remove oxidation just as an eraser eliminates pencil marks. Specialty companies sell abrasive sticks but a simple white eraser is just as effective. The eyelet in **Figure 4-8** shows the difference in oxidation before and after using a white eraser. Using an abrasive stick as a cleaner is not the last cleaning to be done on the part. You must still use the appropriate solvent to meet Class 3 standards.

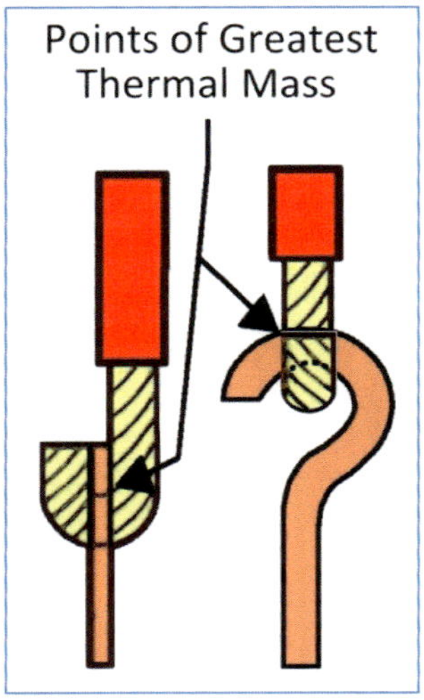

The soldering iron (with a small amount of solder for the heat bridge) is applied to the largest thermal mass (shown by the arrows in **Figure 4-7**) to be soldered. Feed solder into the component when the flux has activated. Once enough solder has been added, wipe the iron off the part; slide it along the wire and eyelet towards the insulation. Don't pull the iron off of the work as solder follows heat of

Points of Greatest Thermal Mass

Figure 4-7
The points of greatest thermal mass are where the soldering iron must be placed during the connecting process.

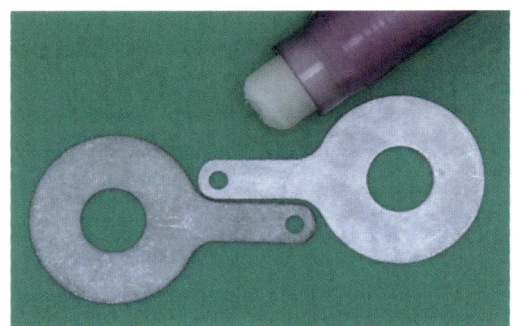

Figure 4-8
Using an abrasive stick (white eraser) to remove heavy oxidation. You can see the improvement in the part to the right.

Figure 4-9
The iron was *pulled* off the eyelet and left a sharp spike of solder (not to mention flux residue and an untinned wire).

the iron (as seen in **Figure 4-9**). If you pull the iron off, you may end up with a solder spike. Slide the iron along the connector towards the insulation. A solder gap of one to two insulation gaps or 1.5mm (use whichever is larger) should be maintained between the component and the insulation. These standards for solder gaps derive from J-STD-001D. In addition, the wire must terminate when it is even with the top of the component. The wire should not extend beyond the part.

It is in the eyelet and hook that you will see a **Fillet** for the first time. All solder joints, with the exception of tinned wire, should have a concave fillet. The fillets have been highlighted **Figure 4-10**. You can see how the fillet has a curve and that it is smooth. This is an indicator that the proper amount of solder was used. Just as in wire tinning, you should still be able to see the original lay of the strands on the soldered wire.

As a rule, the hole that the wire goes through in an eyelet should remain free of solder. However, if the inserted wire is large relative to the hole, the eyelet may be completely filled. A large wire would be one that, when inserted into the hole, does not have room to move side to side.

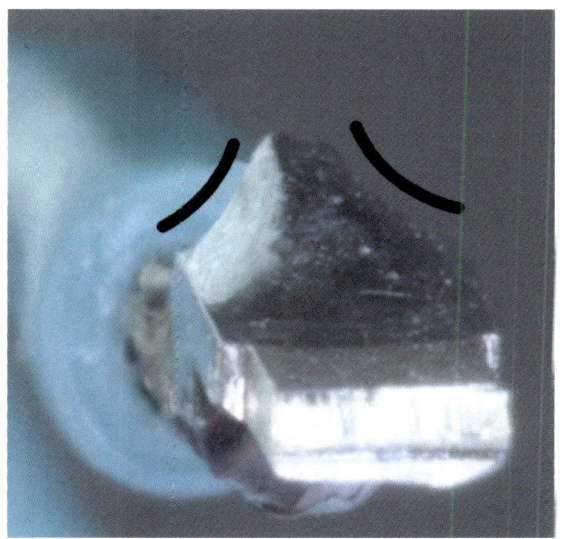

Figure 4-10
Here you can see an eyelet that has been properly soldered. Fillets, highlighted by the black arcs, can be seen on both sides of the soldered connection.

The Turret

The turret is another way of connecting wires to a printed circuit board. A turret with a single wire attached is seen in **Figure 4-11**. This connection is one of the larger masses you may have to solder. Because of

Figure 4-11
A turret with a single wire placed correctly for soldering.

Figure 4-12
This is about as bad as a connection can get. The joint has been severely disturbed while the solder was in the plastic state.

Figure 4-13
Round noses (chain pliers) are used to form the leads which are placed on turrets, posts, pins, etc.

1. Apply iron to point of greatest mass
2. "Paint" solder
3. Wipe iron of work

Figure 4-14
The steps to properly solder a turret. The area where the iron is approaching is the point of greatest thermal mass.

this mass, it can be difficult to work with, particularly when two or more wires must be connected. The large mass keeps the terminal warm for quite some time. During this time, it is imperative that the wire not move until the solder has returned to a solid state. The result of wire movement during the plastic state leads to the disturbed joint of **Figure 4-12**. As always, strip the wire and tin as needed. Clean everything with the correct solvent. Always place wires on the bottom of the turret first and move up as needed. Placing the wire on the bottom lessens mechanical stress to the board, if the turret should inadvertently be subjected to any stress. Next, you will need to make a loop in the wire using round nose pliers (**Figure 4-13**). Select the appropriate diameter for the loop along the pliers' length and test fit to the turret. Make sure to leave the correct solder gap between the turret and the insulation and make sure that the wire wraps 180° or more around the turret. If you are using a small gauge wire, say 30 AWG (0.25mm) or smaller, you must wrap the wire around the post at least once but no more than three times.

Soldering a turret requires that you "paint" the solder on. Flux the turret and apply the iron to the point of greatest mass as shown in **Figure 4-14**. Once the flux has activated, add solder to the area where the iron is then move to the second point and wipe more solder onto the base of the fixture. When sufficient solder has been added, wipe the iron off the turret and do not let the wire move. Any place that the wire contacts the turret must be soldered and exhibit a fillet. It may take up to 20 seconds for the solder to return to its solid state and the turret will remain hot for some time after this. Obviously, you should not touch it during this time!

Several wires can be attached to a turret connector. Again, start at the base and move up (**Figure 4-15**). Always place the largest gauge wire at the bottom. All leads will have to be soldered simultaneously since doing one at a time will cause the entire turret to reach melt temperature and desolder the wire you already soldered. Multiple wires must also exhibit the correct solder gaps and must be parallel to one another.

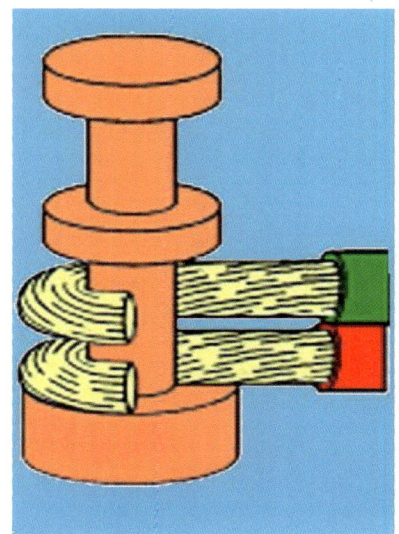

Figure 4-15
Leads should always be attached as low as possible.

Figure 4-16
A properly soldered turret. Note the presence of the fillet.

Figure 4-17
Multiple wires must be soldered simultaneously.

All areas of contact between wires and terminal must have solder on them, have a visible fillet, and the wires original twist should be visible. **Figure 4-16** and **Figure 4-17** show what a proper connection should look like. If viewed obliquely, the connection should be smooth and shiny with evidence of good wetting.

The Bifurcated Connector

The bifurcated fitting offers the most flexibility for making connections. Wires can be routed from the top, bottom,

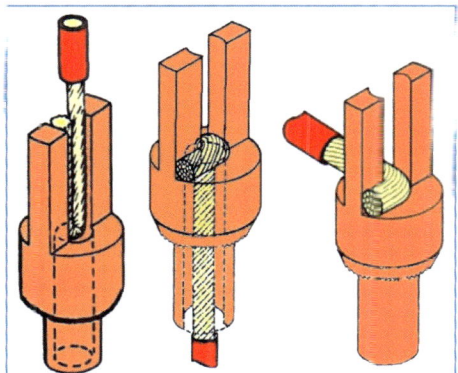

Figure 4-18
Three entry methods for a bifurcated connector.

or sides of the device. **Figure 4-18** shows each of these variants. All three routings have some common characteristics: 1) the wire termination ends flush with the terminal, 2) there is a solder gap between the component and the insulation and, 3) anywhere the wire is adjacent to the terminal, it is in full contact with that terminal.

Look at **Figure 4-19** and you will see what the end product should look like. Solder is present at all contact points, the wire strands are still visible, and the joint is clean. Notice the presence of a fillet - the bottom routed connector shows the preferred amount of solder, the side route exhibits the minimum amount of acceptable solder. As with the turret, the largest gage wire is always on the bottom.

The wire, when side routed, must go around at least one corner of the terminal, it may go around 360°. A wire of 20AWG (0.8mm) or larger *may* be routed straight through because of its large diameter.

Figure 4-19
All of these terminals are acceptable. They have the minimally acceptable amount of solder.

Soldering the bifurcated connector is similar to the turret. Find the point of greatest mass and apply heat there (indicated by the arrows in **Figure 4-19**). On a bifurcated fixture this will generally be the intersection of the wire and the base. As always, make sure the parts are clean and have been fluxed properly. Once you see the flux activate and boil away, paint solder on as required and wipe the iron off the part. Make sure the leads are stationary during this time; any motion may lead to a disturbed joint.

Cups

The cup type terminal still finds much use in the electronics industry, particularly in bulkhead connectors such as the one shown in **Figure 4-20.** The cups can be densely packed and are easier to work on than turret and bifurcated connectors. The greatest problem is preventing solder from running out of the cup or bridging between two cups. Because the cup often fits into a connector with a ridge to lock it into place, solder on this ridge could cause the cup to back out or not seat in its connector. The result is an electrically open circuit!

Figure 4-20
This shell has 10 pins; some can have almost a hundred.

Figure 4-21
A *preform* being cut for a cup. The size of preform must be adjusted for the size of the cup.

The first thing to do with a cup, other than clean it, is to fit the wire. A tinned wire is inserted into the cup and the end trimmed until it meets the 1 to 2 wire gap standard Next, a *preform* may need to be made. A preform is a piece of twisted and trimmed solder that is placed into the hole of the cup (**Figure 4-21**). This is the solder which will connect the wire to the cup. A preform is used in lieu of a solid piece of solder because the preform places all of the solder *inside* the cup. A single piece of solder can be so long that, when the solder begins to melt, the upper portion of solder falls outside of the cup.

Once the cup is heated and the solder melts, inserting the wire should not cause the solder to overflow. The additional twists also add more flux to the cup improving wetting. Overflow, should it occur, is easiest to remove with solder wick to which some extra flux has been added. Unfortunately, the cup will have to be re-soldered (from the beginning) if solder spills out of the hole.

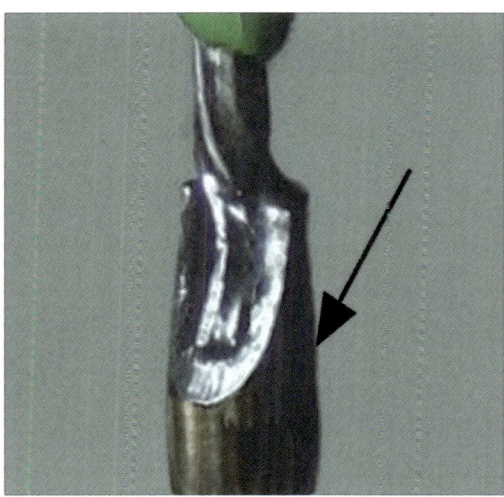

Figure 4-22
The soldering iron should be applied to the cup as indicated by the arrow.

The cup is one of the few solder connections in which you do not add solder to the iron. You don't want to get solder on the side of the cup. If there is solder on the iron, it will transfer to the cup.

The heat bridge for the cup is produced by flux. The iron is placed n contact with the back of the cup (point shown by arrow in **Figure 4-22**). When the preform melts, insert the wire (make sure it bottoms in the cup). Once in the cup, rock wire toward the front and return it to vertical. You should do this two or three time. The motion causes any trapped gasses in the preform to be expelled. Failure to do this could cause pin holes to form because of the effect of *out-gassing*.

The cup should appear like the one in **Figure 4-22**. No solder has flowed out of the cup. The original twist of the wire is still visible and the solder is smooth and shiny. The solder gap has been maintained. As a minimum, 75% of the cup must be filled and proper wetting observed. Some cups have a small inspection hole at the base of the wire socket; if this inspection hole is present on your cup, the solder must be visible in the hole and the solder may extend slightly beyond the hole.

Cups can be also be soldered using ***Resistive Tweezers*** (**Figure 4-24**). These devices have prongs that can be closed on to the cup. The tweezers create a current path between the prongs through the cup to be soldered. The current flow, which is adjusted to a proper value at a soldering station and actuated with a foot pedal or finger switch, heats the cup and causes the solder pre-form to melt. Preparation of the cup for the soldering process is the same for resistive tweezers as for soldering irons. The tweezers are a bonus in cramped areas with several cups. They will heat only the cup they are in contact with and fit into confined areas. Their one shortfall is arcing. This is a technician induced problem, however. To use tweezers correctly, apply the tweezers to the cup, activate the switch, let the solder melt, deactivate the switch, and remove the tweezers

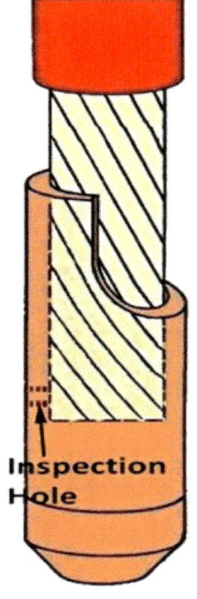

Inspection Hole

Figure 4-23

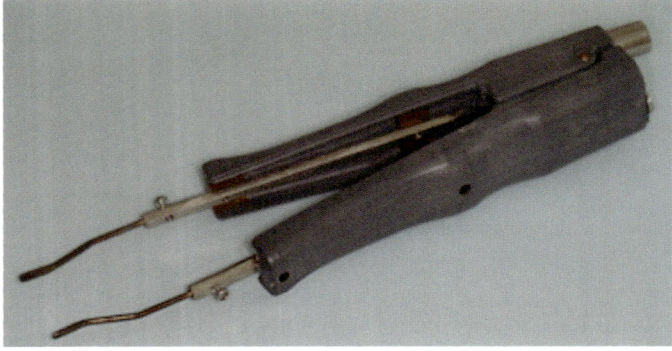

Figure 4-24
One of the many resistive tweezers available to the technician.

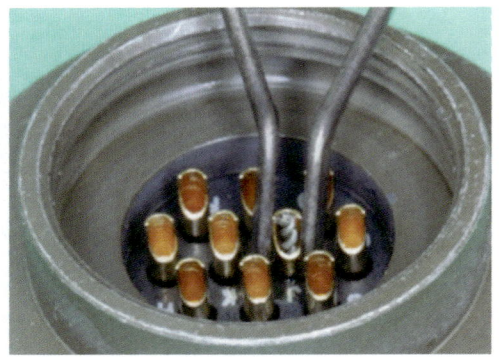

Figure 4-25
A cup ready to be soldered into place.

leads. The arcing occurs when the technician activates the tweezers *before* contacting the cup or leaving them activated when removing them from the cup. The slight gap, which occurs when opening or closing the tweezers, causes a voltage to jump from the tweezers to the cup, possibly damaging the cups plating.

Connectors with Gold Plating

Gold plating is used to protect connectors where deterioration of the connector might occur. Working with such connectors is not difficult but some precautions must be observed. Any connector that has less than a 2.5 micrometers (μm) must be tinned once before it is soldered. Gold plating equa to or greater than 2.5μm has to be tinned twice. Tinning helps prevent brittle intermetallic bonds because tinning removes the gold plating. A direct tin/gold connection would be weak so the gold must be allowed to diffuse at the contact points. This is the reason for tinning. It is extremely important that only the point to be soldered is tinned! Tin the entire connector and you have defeated the purpose of gold plating.

Soldering Wire Splices

The splices covered in Chapter 3 should also be soldered to assure a permanent connection. Additionally, they should be covered to protect them against shorting and potential environmental degradation. Some methods of covering wires will be considered shortly.

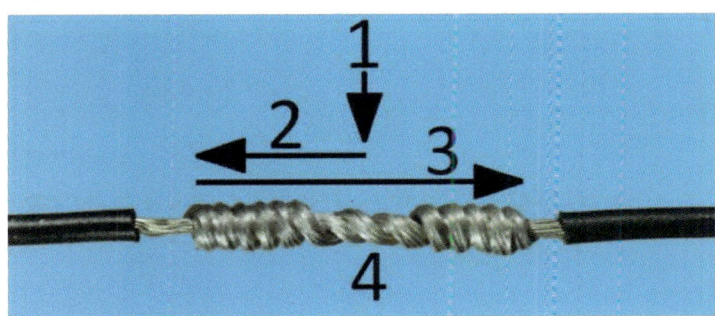

Figure 4-26
To solder Western Union Splice, begin at 1; slide the iron to point two, then to point 3. You will have to adjust the amount of solder fed into the splice because you should cover all of the wires but still be able to see the strands when finished.

Soldering the Western Union Splice will be discussed first. Once you have cleaned the wire, slide on a tube of heat shrink (see **Sealing the Western Union Splice**) and apply flux carefully. Applying too much flux can cause it to wick under the insulation. Solder will travel as far as the flux has gone and this leads to unwanted capillary action. **Figure 4-26** shows the movement of the soldering iron during the process. Touch the iron to point 1 to activate the flux (don't forget a small amount of solder on the iron for the heat bridge). As you draw the iron to point 2, continue to add solder as needed. Move the iron only as close to the insulation as needed for wetting. Remember, the solder will follow the heat. Slide the iron to the opposite side and continue to add solder until you get to point 4. At this point, use the heat of the iron to draw the solder the remaining distance to the end of the splice. You may have to adjust this last step for your particular "touch."

When completed, the wire strands should exhibit their original lay. The solder should be shiny, smooth, and exhibit good wetting. Many technicians want to move too fast while soldering this splice so cold solder spots are common. There also seems to be a tendency to apply solder too quickly. It may appear to wick nicely into the wire but when the soldering iron is removed and the opposite side of the wire is inspected, a large amount of solder has collected. The only way to prevent this … practice.

Sealing the Western Union Splice or any other straight connection is best accomplished with *heat shrink*. Heat shrink is a type of organic compound (plastic) that is tube shaped. When heated, it decreases in size, it shrinks. The decrease is in diameter more than length. A typical heat shrink might say it has a 2:1 shrink ratio. If the heat shrink began as a ¼" (6mm) tube, it would shrink to 1/8" (3mm). It would not shrink as much along its length, 15% shrinkage might be a normal specification. If the ¼" (6mm) tube from the last example were two inches (5cm) long before heating, it would decrease to about 1 3/4" (4.5cm) in length. Typical shrink ratios range from 2:1 to 4:1.

Heat shrink comes in many materials, colors, and sizes. Heat shrink can also be selected for its shrink ratio (how much it shrinks when the correct amount of heat is applied), chemical resistance, moisture resistance, temperature capability, abrasion resistance, voltage rating, and wall thickness.

Polyolefin and PVC are general use heat shrinks. They have relatively low activation temperatures but have low temperature resistance as well. PVC heat shrink can easily have holes burned into it with a heat gun. It also tends to get stiff at low temperatures.

Teflon (PTFE) and Fluorinated Ethylene Propylene (FEP) heat shrinks are more temperature and chemical resistant than polyolefin and PVC but cost more. FEP has less chemical resistance and is less costly than Teflon. It will withstand temperatures as high as 500° F (260°). Because of its high shrink temperature, it is usually applied in temperature controlled ovens. PTFE is the most difficult heat shrink to work with.

Polyvinylidene Fluoride (PVDF) has the ability to withstand continuous high temperatures to about 300°F (149°C). PVDF is more commonly known as Kynar, the same material used as an electrical insulator. The temperature required to shrink PVDF is roughly 605°F (318°C).

Neoprene heat shrink is also available. It is reasonably priced, offers good chemical and good abrasion resistance. It works in temperatures up to 245° F (118°C) and the shrink temperature is approximately 265° F (129°C).

When applying heat shrink, use a heat gun. Do not use an open flame of any type. Heat guns are designed to provide a controllable temperature range and air velocity. Open flame lighters subject the heat shrink and the electrical insulation to temperatures that exceed the specifications of both. Heat guns can be dedicated devices costing several hundred dollars to simple two speed devices as shown in the image (along with a small sampling of heat shrink). The costlier versions have much better temperature control and interchangeable nozzles so heat can be applied very accurately. Whatever device you use, be safe. The air exhaust temperatures from a heat gun can easily melt solder; the solder splice used in Chapter 2 was melted using such a heat gun!

Our Western Union Splice is now ready for heat shrink. Maintenance people occasionally run into a problem here. They have spliced and soldered the connection before sliding the tube of heat shrink onto the wire. If you do this, your only option is to break the newly joined connection and start again, Not convenient at all. So, slide the heat shrink onto the wire before splicing it. Select the correct material based on whatever criteria have been established by you or your firm. If you are using a 2:1 ratio material, select the diameter of heat shrink that allows you to slide it over the, heat shrink should never be forced over a splice. You always want the heat shrink to extend past each end of the connection by about 3/8" (9.5mm) This will leave ¼" (6mm) protection after shrinkage. Once positioned, slowly heat the area with the gun. Keep it in motion and make sure to heat all sides. Within a few seconds, you should see the shrink contract. Once the heat shrink stops contracting, stop heating it.

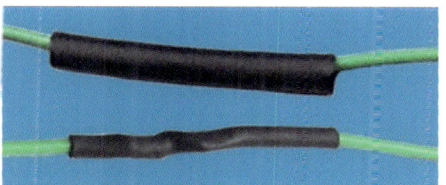

Figure 4-27
A heat gun with several sizes of heat shrink.

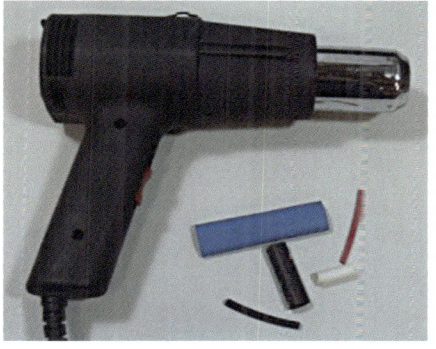

Figure 4-28
Heat shrink before and after heating.

Resist the urge to touch the heat shrink. It or the connection underneath could still be hot. The splice is now protected. You can see in **Figure 4-28** how effectively the wire has been sealed. You now also understand why it was important that the ends of the Western Union splice not stand up. They could quite easily puncture the heat shrink.

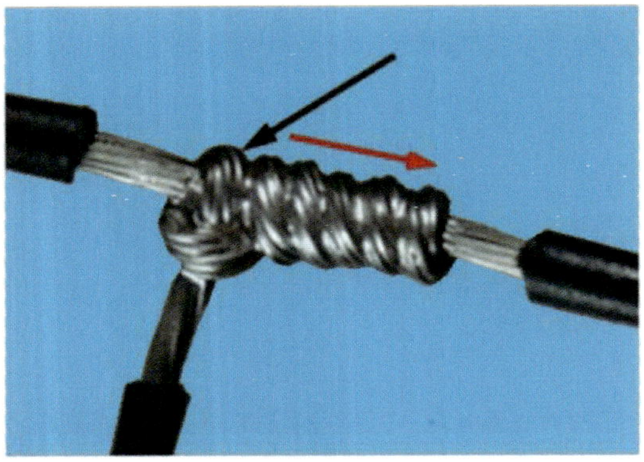

Figure 4-29
Begin soldering a tap splice at the arrow marked 1, then slide the iron in the direction of arrow 2.

Soldering the Tap Splice is the same as soldering the Western Union. Clean and flux as appropriate and then apply the soldering iron at the point of greatest mass (place the iron on the back of the connection as indicated by the black arrow in **Figure 4-29**. Feed solder onto the wire and allow it to be drawn in via capillary action. Move to the opposite end of the wire (red arrow) and pull the iron off. The entire process should take about five to seven seconds depending on the wire gauge. As in all soldering, a little practice will let you know when to stop adding solder and how quickly to move. Everyone works a little differently. The wire strands should be discernible, the solder smooth, and exhibit good wetting. None of the solder should have wicked under the insulation.

Sealing a Tap Splice is more challenging that sealing a Western Union. Heat shrink cannot be used (*if* soldered as seen in **Figure 4-29**) because of the perpendicular connection of the wires. If heat shrink were used, the tap would have to be bent parallel to the original wire. When the heat shrink is activated, it would shrink and place considerable stress on the bent wire. The small radius of the bend would be compressed and strands could break. If the tap was bent parallel to the original run *before* soldering, heat shrink can be used.

Splicing tapes are a good choice for this problem. This is not the electrical tape you buy at a hardware store! Properly, the tape to use is a linerless rubber splicing tape. They are flame retardant and able to provide insulation up to several thousand volts. It should also allow any heat that may inadvertently build up to be carried away. When selecting a splice tape, try using a narrow roll, say 3/8" (9.5mm) to 1/2" (12.5mm). This allows tighter fits on the small splices sometimes found in electronics. If you are working with larger gages, 16AWG (1.3mm) or greater, use ¾" (19mm) tape.

Applying the tape is simple. Remove as much as needed from the roll. With the tacky side up, stretch the tape and then wrap it around the connection making sure to overlap at least half the width of the tape. The tape will adhere to itself and form and excellent barrier physically and electrically. Making this splice look good is tough but be patient. Why make it look good? A technician is judged on many things. One of the criteria is the appearance of the work they do.

Different Solder –

Different Appearance

The prior chapter discussed the chemical characteristics of various solders in detail. Since this chapter covers usage, it is an opportune time to actually see the differences inherent in each type. **Figure 4-30** shows the appearance of different solder compositions.

Three types of solder were used. Notice that all three types of solder created a proper filet and completely wet the pad. The pin marked **A** was done with Sn97Cu.2Ag.8Sb2, a lead free solder. Notice how granular this connection appears compared to the

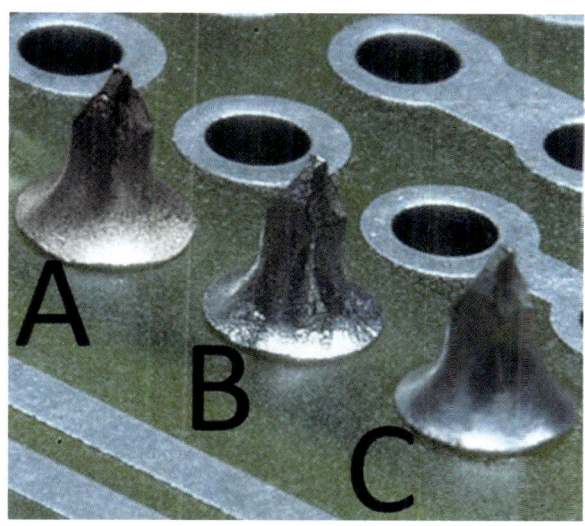

Figure 4-30
Connection A is of tin, copper, silver, and antimony. Connection B is tin, lead, and silver. Connection C is tin and lead.

others. This is quite normal and acceptable. Joint **B** was done with Sn62Pb36Ag2. This is not a lead free connection but it does exhibit some slight granularity. Connection **C** was done with Sn63Pb37, a lead containing solder with eutectic characteristics. Its surface is completely smooth, a trait of lead and tin only solders. Dwell time for all three is similar. The eutectic solder was applied with the iron set at 610°F (321°C). The temperature of the soldering iron was adjusted 10° to 20° F (6° to 12°C) higher for the lead free solder. These values are not fixed. Lead free solders may need a more active flux because their wetting ability is inferior to lead containing solders.

Chapter 5

Printed Circuit Boards

The *printed circuit board* (PCB) is literally the nervous system on which electronics are built. The board acts as the nerves of a circuit by interconnecting all the components. The PCB has seen extensive change over its century of existence. This chapter will cover some of those changes, both in usage and manufacture, and establish a foundation for topics to be discussed in the following chapters.

In the Beginning

The concept of using a platform to mount electrical components on has a long history. As far back as the 1850's, interconnection technologies existed. In this first case, metal rods were attached to a board and connected to components via wires. Albert Hanson suggested cutting traces on copper foil and then gluing the traces to a substrate in 1903. Another patent was issued in 1913 to Arthur Berry. He also used a copper foil but instead of cutting traces, preferred etching the traces on the board with a solvent. In 1925, a patent was given to Charles Ducas. His plan was to have circuits placed on a board using conductive ink forced through a stencil. Conductive inks are still used, primarily by hobbyists.

You may have heard of *breadboarding*. This pre-PCB design method was used by amateurs and inventors to create circuits. A wooden breadboard had nails driven into it. These nails acted as binding posts for component leads. Breadboards are an adaptation of the 1850's metal rod technique. The breadboard (also known as a proto-board) we use in electronics experimentation today derives its name from this method of interconnection. Back to the original breadboard. Making connections with nails on a board with wires between them, connecting component to component, was known as *point-to-point*

wiring. This was not at all satisfactory for connecting large numbers of components and was very susceptible to electrical noise. However, because of its low cost, point-to-point wiring was used in lieu of the more expensive process of placing conductive traces on an insulated board.

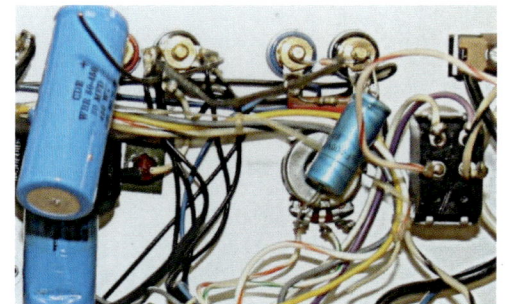

A practical method for creating a circuit board was patented by the Austrian scientist Paul Eisler. He suggested using copper on a glass insulated base and applying photo-etching to create the circuit paths. Although the idea was

Figure 5-1
An example of *Point-to-Point* wiring.

patented in 1943, it initially saw no commercial use. Its first application was in fuses used by J.S. Army Ordnance Bureau. The fuses were proximity devices originally designed to shoot down German V1's (the *Buzz Bomb*) during latter half of World War II. They were soon adapted to all types of anti-aircraft ordnance. Eisler is reputed to have made his first printed circuit board in Great Britain in the 1930's. The Bendix Corporation had also developed the same process and it fought Eisler over usage rights until 1963. Eisler lost. The idea of Eisler and the Bendix developers is essentially what we have today or our most commonly used printed circuit boards.

Commercial use of PCB's did not start until the 1950's. This coincided with the invention of the transistor. The 1950's saw the advent of through-hole technology – drilling holes through the board to mount components. This allowed for a far greater density of components. Cordwood construction was also used during this period. Two boards were placed a short distance apart and components were mounted from one board to the other (see **Figure 5-2**). The components were stacked like cords of wood. Eventually, integrated circuits (IC's) forced manufacturers to develop multi layer boards. These are boards that had a layer not just on the front and back of a board, but had copper sandwiched between the glass (or phenol) insulating layers.

Figure 5-2
A board using cordwood construction.

Board Types

The majority of circuit boards currently manufactured are an epoxy and fiberglass composite Printed circuit boards are now resilient and relatively inexpensive products. Some companies still use the older Phenolic materials because they are cheap. Phenolic boards use phenol and an aldehyde to form a bonding agent that is then placed on layers of cotton paper. Standard thicknesses for all of these boards are 0.020" (0.5mm), 0.031" (0.787mm), and 0.060" (1.5mm). Circuit boards are given grades based on the materials used in their construction (see **Table 5-1**). For instance, **FR-1** and **FR-2** (the FR stands for Flame Retardant) are Phenolic boards. FR-1 is a paper and Phenolic material. As such, it is not stable in high moisture environments. FR-2 uses cotton paper and has better moisture tolerance than FR-1. FR-1 and 2 are used in single sided applications. Copper traces are placed on one side only. More on this later. One of the trade names for FR-2 is Bakelite – a product in common use in the 1950's.

| **Figure 5-3** | **Figure 5-4** |
| FR-1 Board | FR-2 Board made of Bakelite |

Almost all boards used in electronics today are **FR-4** which is a woven glass and epoxy blend. The fiberglass/epoxy combination has good electrical characteristics and is more flame retardant than FR-1, -2, and -3. If it does ignite, it is self extinguishing. It can also be multilayered; up to eight thin sheets of copper can be sandwiched between layers of FR-4. FR-4 is also very easy to machine meaning that it can be sheared to the proper size and drilled easily and cleanly. It is stable up to about 245° F (118°C) stability varies with the thickness of the board).

Boards made of G-10 or G-11 material have the electrical characteristics of FR-3 but a higher temperature rating. The **CEM-3** grade (CEM = *C*omposite *E*poxy *M*aterial) of PCB is very similar to FR-4. CEM-3 boards are woven glass placed over a non-woven core with epoxy as the bonding agent.

FR-5 is closely related to FR-4 as it is also a fiberglass and epoxy construct. It is more resistant to chemicals and moisture than FR-4. It also tends to retain its flexibility at elevated temperatures.

Other materials are found in high-end products. Polyimide board substrates are one of these. The polyimide board is found in leading edge computers. Teflon boards have been used and are found in telecommunication and aviation

Grade	Material
CEM-1	Cotton paper/epoxy
CEM-2	Cotton paper/epoxy
CEM-3	Woven glass/epoxy
CEM-4	Woven glass/epoxy
CEM-5	Woven glass/polyester
FR-1	Paper/Phenolic
FR-2	Cotton paper/Phenolic
FR-3	Cotton paper/epoxy
FR-4	Woven glass/epoxy
FR-5	Woven glass/epoxy
FR-6	Woven matte glass/epoxy
G-10	Glass cloth(matte and woven)/epoxy
G-11	Glass cloth(matte and woven)/epoxy
Getek®	Polyphenylene Oxide/Epoxy resin, reinforced with electrical matte glass.

Table 5-1
These are just a few of the circuit board materials available. Many companies manufacture proprietary substrates which are not listed here.

devices. Even ceramics have come into use as circuit board substrates. The ceramics excel in high frequency applications. Getek® boards are used in circuits requiring high speed and low dielectric loss

Making the Connection

Making a connection between components requires a conductive path. All of the PCB types listed previously can be clad in copper by the manufacturer. Most often, the copper is in the form of a sheet that is glued to the substrate. The board may be classified as a 1/2-ounce (14g), 1-ounce(28g), or 2-ounce (56g) board. This indicates the weight of copper per square foot. A one ounce board has one ounce of copper per square foot. The more copper, the higher the current carrying capability of the board. It is important to select not only the correct substrate for the application but also the correct copper weigh: don't use a two ounce board when a ½ (14g) ounce board will do, a two-ounce board (56g) may cost four times as much as 1/2 (14g) ounce board.

Another method for affixing copper to the board is to use electrolysis. This transfers copper only to the areas where it is needed. The increasing cost of Cu is making the latter method preferable. These and other fabrication methods will be discussed at length later in this chapter.

| Figure 5-5 |
| A single-sided board. |

| Figure 5-6 |
| An example of pads and traces. |

Single Side Boards

PCB's can be single sided (copper on one side only), double sided (copper on both sides), or multi-layered (16 or more layers sandwiched between sheets of substrate). A single sided board is shown in **Figure 5-5**. The board has already been etched and was sectioned to show the single side. The pads were not drilled. The substrate on this board is FR-4. **Figure 5-6** shows the types of traces that may be seen on a single sided board. Any copper to which a component may be attached is called a **Pad**. The pad to the left of **Figure 5-6** is for a through-hole component, a resistor with an axial lead for example. The square pad to the right would be used to attach a surface mount component. The connection between the two pads is a **Trace** (also called a track). Its thickness (based on the boards copper weight) and width determine the amount of current that can safely be carried.

The single sided board has a serious shortcoming; the complexity of the circuit is extremely limited. Imagine a device with six pins (as seen in **Figure 5-7**). Try to connect A to 1, 2, and 3. Repeat the procedure with each letter. The only rule is that you may never cross another connecting line. This dilemma faces anyone trying to make multiple connections on a single sided PCB. The solution is to add another layer of copper. This gives us the double sided board.

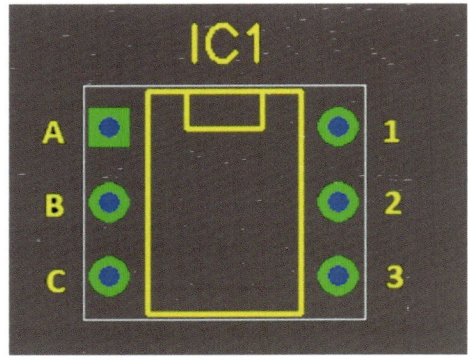

Figure 5-7
Try connecting every letter to every number without crossing any lines.

Double Sided Boards

The double sided board has the usual one-half to two ounces of copper but has it on each side of the board. This greatly increases the number of possible paths that a trace can be routed along. **Figure 5-8** shows a double sided board that has been sectioned. It is easy to see that the copper goes through the hole that was cut into the board. Any connection like this is called a ***Plated Through-Hole (PTH)***. Sometimes a connection must be made from one side of the board to the other without having a component installed through it, i.e. there is no hole through the board. Such a connection through the board is called a ***Via*** (see **Figure 5-9**).

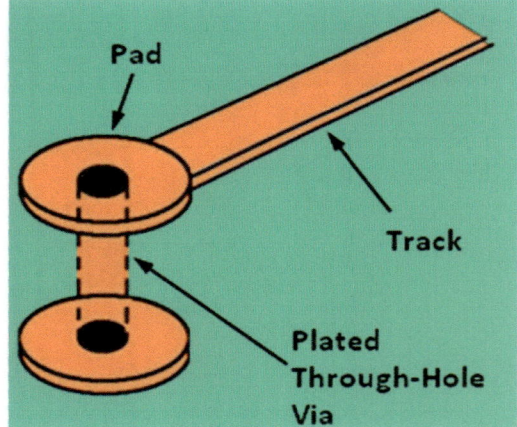

Figure 5-8
A double-sided board with plated through-holes.

Figure 5-9
Tracks, pads, and vias make up a double-sided board.

Multi-Layer Boards

Ever increasing complexity of electronic circuits and the accompanying component miniaturization has lead to the need for multilayer boards. Boards with as many as 48 layers are commercially produced. Multilayer boards can be made in several ways. If a 4-layer board is desired, a single layer board can be placed on either side of a two layer board. Another option is to use a pair of two layer boards and separate them by a *prepreg*. The prepreg is a layer of FR-4 that has not been completely cured and is not clad in Cu.

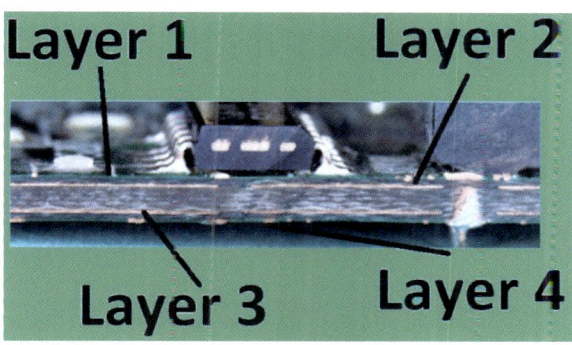

Figure 5-10
This board was cut to show the layers inside. Note that the chips and pins were also sectioned. The board was from an early 1990's desktop computer.

The tracks on some boards have become so fine that a static discharge can completely blow the track off the board. Electrostatic Discharge (ESD) mitigation has become an imperative for this reason (see Appendix A for more information on ESD).

Multilayer boards are typically laid out in specific sequences. The four-layer board seen in **Figure 5-10** has the signal layers on the top and bottom of the substrate; sandwiched between these are the power and ground layers. A seven layer board might have a signal layer first, then a power layer, another signal layer, a spacer layer, a third signal layer, a ground plane, and finally the fourth signal layer. The order of layout affects the sensitivity of the board to electrical noise. Not every pin on the integrated circuit (IC) needs to be connected to the ground or power plane. An area void of copper is left wherever signal pins pass through ground and power levels.

The multilayer board can have electrical contacts between every layer. This board presents two new problems to the technician; the *buried via* and the *blind via* (**Figure 5-11**). The buried via is *hidden entirely between the layers of the board. The*

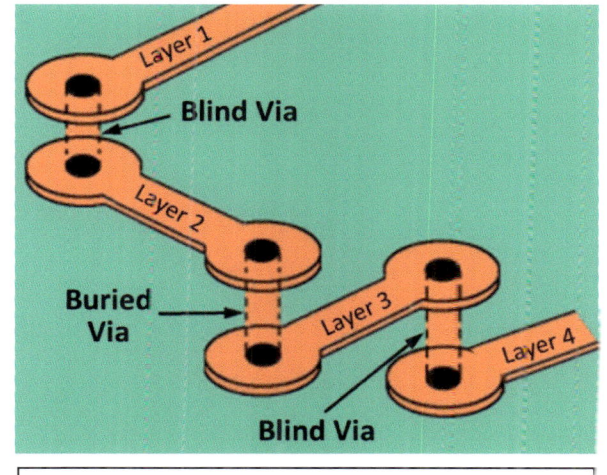

Figure 5-11

blind via has one connection that can be seen from the outside of the board and is attached to a hidden connection on one of the inner layers. Problems arise when one of these vias fails because of the traces inaccessibility.

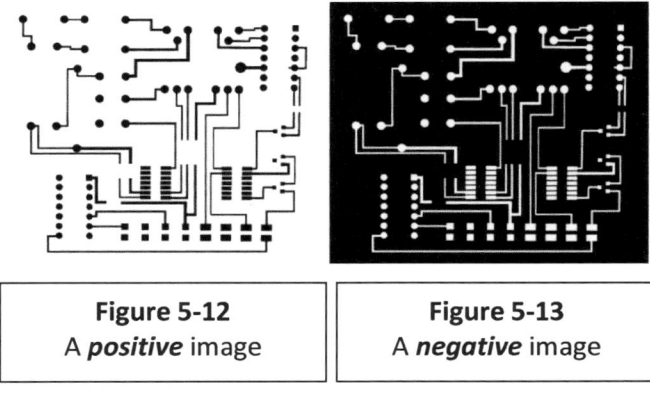

Figure 5-12	**Figure 5-13**
A *positive* image	A *negative* image

The Artwork

Computer-Aided Design (CAD) eases the development of PCB's considerably. Not so long ago, designers would use templates and stencils to draw the circuits under design. Once the design was approved, the artwork would be photographed and a positive or negative image made. Both images would be printed on a clear photographic sheet of acetate film. The white areas on the images would be clear on the acetate sheets. **Figure 5-12** shows the original design; **Figure 5-13** is the negative image. Both images can be used to make circuit boards. A board used with **Figure 5-12** would have the exposed resist (resist is covered in **Photo-Developing** on the following pages) softened and leave the tracks and pads in their original state. With **Figure 5-13**, the exposed resist would harden and all the unexposed areas would be removed.

Many programs are available to create circuit images. Some professional versions can cost several thousand dollars. A few design programs are available as freeware and are quite good. They can generate positive and negative images as well as *Gerber files* needed for milling the PCB (see **Milling** under **Subtractive Methods**). Stencils and templates are still available.

Making a Circuit Board

Once the artwork or pattern for the printed circuit board has been made and approved, the image can be transferred to the board. There are several ways to make a PCB. These methods fall into two categories, additive methods and subtractive methods. The subtractive approaches are still the most prevalent and will be discussed first.

Subtractive Methods

The subtractive process, as its name implies, subtracts or removes material. The principal uses of subtractive methods are in single or double sided boards. Multi-layer board manufacturing is easier with an additive process when boards of four or more layers are required.

Photo-Developing

The circuit image must be applied to the board in some way. The most prevalent is to transfer the image through the artwork onto a material called **resist** or **photo-polymer**. This material is sensitive to *ultra-violet* (UV) light. It can be applied by the manufacturer (a **pre-sensitized** board) or added immediately before the board is to be used. The board is considered **non-sensitized** in the later case. When the resist is exposed to UV through the clear areas of the artwork (now on the acetate film), the resist either hardens or degrades. If ultraviolet light can cure the resist (harden it), this method is called a *negative resist process*. If the ultraviolet degrades the resist, you have a *positive resist process*. Hardened resist protects the area under it. Softened resist allows the area beneath it to be removed.

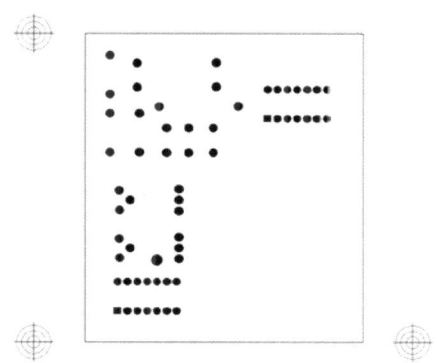

Figure 5-14
The "targets" (left top and bottom, right bottom) are the **registration** marks.

The device that contains the UV light source and holds the artwork in place is called an Exposure Frame. This is essentially a box with ultra-violet lights fixed to the bottom and, occasionally, the lid. In this way, the ultra-violet light would expose one or, if the PCB is double-sided, two sides of a board. Creating double sided boards requires careful alignment of two images. Imagine a two sided board that has through-hole connections. A misalignment can cause opens and shorts, making the board useless. **Registration Marks** are added so that both sides of the artwork can be accurately aligned. You can see three registration marks in **Figure 5-14**.

Once the image is on the resist, you must remove the degraded or uncured resist by applying a chemical. This **Develops** the board. A typical developer might be **sodium carbonate** (washing soda) which is a base (pH greater than 7). Developing leaves the hardened resist behind and it protects the areas of copper you wish to keep. At the same time it removes all the uncured resist. Developing also stops any further exposure of the resist. Stopping exposure of the board at this point is imperative;

almost all artificial light sources emit some UV. Exposure to sunlight has the same effect. Leave a board out long enough and it will be exposed which makes it useless.

When developed, the board will have a bluish hue over the circuit tracks. The remainder of the board is bare copper. This is called a negative resist process because the UV passes through the clear areas of the mask and hardens the resist it hits. The image in **Figure 5-13** would be used in this process. With the positive resist method, UV passes through the transparent areas and degrades the resist. **Figure 5-12's** mask would be used for this type of procedure.

The next step is to remove the copper you don't need. This is **Etching** of the board and *it is done in an Etchant Tank or Etcher. One common etchant is **sodium persulfate**. Ferric Chloride* is also used and it is still available. However, the use of ferric chloride should be avoided because of its environmental impact. Sodium persulfate is a strong oxidizer that changes the bare copper to copper chloride. The copper chloride washes off the board during the etching process.

Etchers are boxes that contain the sodium persulfate and spray a fine mist of the chemical on the board's surface. The chemical is heated slightly (about 100° F) to make it more effective. A typical etcher might have two pumps, placed on either side of a tank which holds the *etchant*. These pumps drive etchant through perforated tubes. The fluid strikes screens, called *diffusers*, that create a mist to evenly cover the PCBs. The PCBs are suspended in the center of the tank. Full immersion etcher are also used. The PCBs are placed in the etchant which is agitated by air bubbles. After a few seconds of exposure to the etchant, the copper will begin to look dull as the reaction takes place. Only a few minutes are required to remove all of the unwanted copper. Leaving the board in the etchant too long can lead to a serious problem. The etchant can force its way under the hardened resist and remove it. Eliminating the resist leaves the tracks unprotected and subject to removal.

Stripping is the last step in board development. ***Potassium Hydroxide*** (*lye* or *Caustic Potash)* is used to remove the hardened resist. Adding the potassium hydroxide crystals to water produces a strong alkali solution which is corrosive to the hardened resist. If you mix the chemicals, always add the crystals to water, not the other way around. Once mixed, simply place the board into the liquid and let it soak for a minute or two. Stripping solution that has seen considerable use may take longer to work. Agitating the potassium hydroxide solution after letting the board sit in it for a few minutes will cause the hardened resist to float off of the board.

Figure 5-15 shows how the boards look after each step in the boar manufacturing process. The boards are: exposed, developed, etched, and finally stripped.

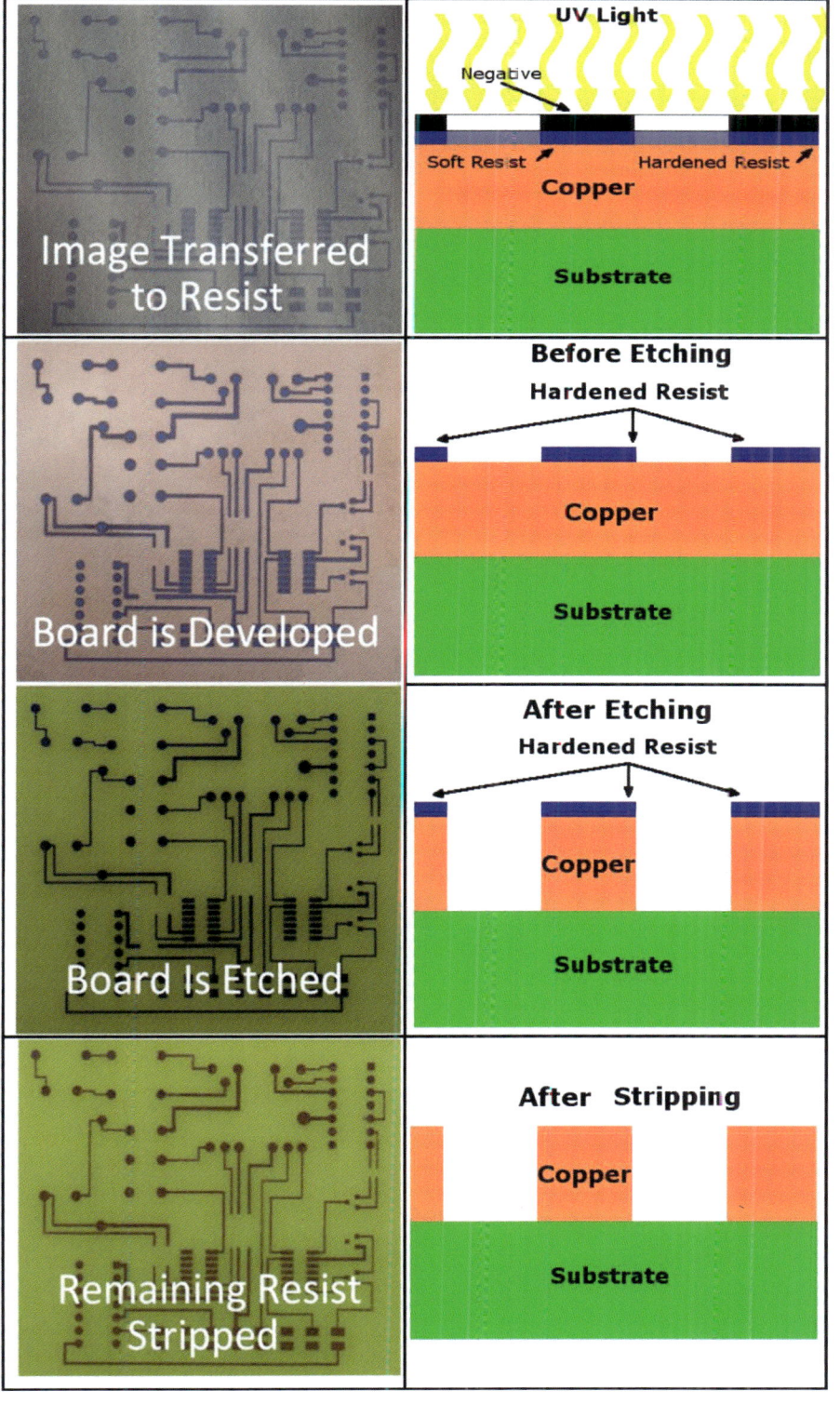

An alternative method to the processes discussed, especially where plated-through holes are required, is to use electrolysis. Before the board is exposed, all of the through-holes are drilled. The board is then exposed and developed. At this point, the board may be placed in a bath of **copper sulphate** and **sulphuric acid**. A voltage is applied to the copper sulphate plates and electrolysis causes the copper to migrate to the board, plating it and the through-holes. The board may then be tinned, again through electrolysis.

Figure 5-15
Here is what you would see as the PCB is created.

If you are mixing any chemicals, take all precautions recommend by the manufacturer and relevant agencies.

The components and traces placed on circuit boards continue to shrink in size. For common electrical systems, the photoengraving process just discussed is quite satisfactory. Several methods exist to produce detailed circuit drawings beyond the capability of photoengraving. Among these are lithography and laser etching.

Lithography uses polymers (short chain molecules connected to form long chains) and metal plates to create the circuit images. Variations of lithography are *microlithography* (features as small as 10μm can be drawn) and *nanolithography* (features as small as 100nm can be drawn). Lithographic images can also be drawn using electron microscopes. This is called electron beam lithography and can reproduce images of a few nanometers.

Laser engraving uses a laser to ablate (remove) materials from a substrate. The laser is moved along the X and Y axis much like a PCB mill (see **Milling**). Accuracy is down to the nanometer range.

Silk Screening

The silk screening process applies resist to the areas to be protected. The screen is placed over the copper clad board and resist is applied to one side. A floodbar (squeegee) is passed over a stencil. A stencil (the silk screen) made of polyester with very fine pores, allows the resist, forced through by the squeegee, to flow on to in selected areas. Silk screening can be automated or

Figure 5-16
All the lettering in this picture was silk screened. You can see how the lettering makes component placement much simpler.

manual. A positive or negative image can be created using this method. Once resist is applied, silk screening follows the steps of a photoengraving process, etching, stripping, etc.

Silk screening may also be used after stripping. Quite a few companies mark the board with component designations (**Figure 5-16**). This makes manual fabrication, and any repairs that may eventually be needed, much easier to accomplish. The lettering helps the assembler or technician locate and properly orient the designated component in much less time than when compared to boards that lack such designations.

Milling

Another of the subtractive processes is PCB milling. A mill capable of moving along X and Y coordinates removes unwanted copper while an X, Y, and Z capable mill can remove unwanted copper and drill all through-holes. **Figure 5-17** shows the orientation of these axes.

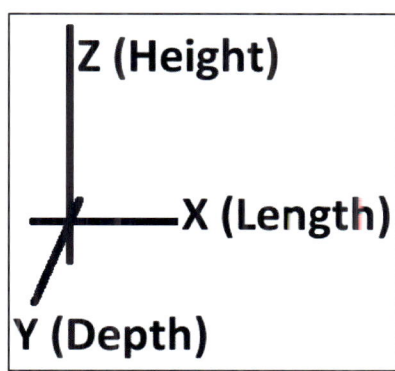

PCB mills, or PCB routers as they are sometimes called, are computer controlled devices. Mills and routers controlled by computers are also known as Computer Numerical Controlled (CNC) machines. The software used to create the circuit board is contained in Computer Aided Design (CAD) software. Most CAD programs will generate a set of files, called *Gerber* files, with these extensions:

Figure 5-17
The three axes that many mills can operate on.

.gbl – the bottom layer of the board

.gbt – the top layer of a board

.gbs and .gbt – the bottom and top resist layers

.gbo and .gbt – the bottom and top board overlays

These files will contain all of the information for a printed circuit manufacturing facility to create the board on a computer controlled mill. The manufacturer loads these files on a computer with the appropriate software. The computer then reads the files and controls the motion of the mill's heads. The files also contain information on which bits to use – a bit to remove copper is not the same as the one that drills the through-holes.

Figure 5-18
A commercially available three-axes mill suitable for prototyping work.

The strength of milling is that the process can be accomplished with fewer steps than photo etching. Developing, etching, stripping, and drilling of the board are unnecessary. Milling is an excellent way of quickly creating a prototype board. More advanced mills can change the cutting tools automatically allowing the board to be created and drilled in fewer operations. Simpler mills will require the bits be changed manually. Plated through-hole boards will still require electrolysis or some other deposition method to create the through-hole connection. The principal drawback is cost. Economy mills will cost a minimum of $10,000. Also, it may be necessary to learn software unique to the manufacturer of that mill and have extra bits on hand as the older ones lose their edge.

Additive Methods

One of the greatest drawbacks of subtractive processes is that much of the copper on a board is wasted. Traces and pads make up only a small fraction of the copper originally applied to the board. This may make additive processes the wave of the future. *Additive processes* selectively apply copper only to the areas that need to be conductive. Additive methods also allow for the creation of very fine traces. Additionally, the additive process lends itself to multi-layered boards quite well.

Additive processes add copper to a bare board. First, a resist is applied to the board. A mask is then laid over the resist and the board is exposed to ultra-violet light. An image such as **Figure 5-13** would be used. The traces are the areas on the board at which the resist is degraded. A suitable solvent is then used to remove the degraded resist on the traces and pads. The board is placed in a copper sulfate solution. This is an electroless plating process that causes the copper to deposit on the bare copper areas of the board through catalytic reduction. Electroless means that no electricity was used in the plating. Finally, the hardened resist is removed.

Tinning

Tinning involves the application of tin or tin/lead to the exposed copper traces and pads. Tinning of PCBs is extremely important. To stop oxidation of bare copper and to prepare all contacts for component mounting, it is absolutely necessary to tin the board. There are several methods available to tin a board.

1) **Wave soldering** – This requires you to place the bare board over a pool of solder. A wave is generated in the solder tank and travels down the length of the pool. As the wave travels, it comes in contact with the bare copper traces and applies a thin layer of solder, effectively tinning the work. The board must be cleaned and appropriate flux applied to the tracks before tinning.

2) **Roller tinning machines** – These devices also have a bath of molten solder but instead of making a wave as in wave soldering, this device has two rollers, one of which applies solder to the board. It is

much like the old washing machines that had a crank to force clothes between two rollers. Instead of removing water, tin is applied. In roller tinning, the board is placed between the rollers. As the board is fed through them, a thin solder layer is applied to the board. The thickness of the tinning is determined by the pressure between the rollers. Cleaning and fluxing of the board must be done before the tinning process commences. Tinning material can be applied to one roller or both.

3) **Chemical plating** – Chemical plating uses an inorganic tin compound like stannous sulphate along with an acid, perhaps sulphuric acid, to create the tin layer on boards. The board is merely dipped into the solution and plating begins almost immediately via chemical reaction. Spraying of the board with the proper solution can also achieve good tinning quality. Many proprietary tinning solutions are on the market and are quite effective. Tinnit and Immersion Tin Ormecon are two examples. As with all chemicals, follow the manufacturer's instructions and dispose of waste chemicals properly. One added word of caution. Tinning uses an acid while some of the board creation methods mentioned previously use oxidizers (base materials). These two DO NOT MIX, keep them well separated!

4) **Solder wick tinning** - The three methods mentioned above are capable of doing a small production run or a run of millions of boards. They are not reasonable solutions for board repair or for a few prototype boards. Simple solder wick (also called desoldering braid) lends itself well to this process. Solder wick is a flat copper mesh in varying widths (0.025" to 0.125" (0.6mm to 3.2mm) are common) that may have dry flux in them. Wick removes excess solder principally through capillary action. The wick is placed on the point to be cleaned of solder; an iron is then placed on top of the wick. This causes the wick to heat first then the solder to melt. Capillary action draws the liquid solder into the wick. The heat from the iron also helps solder movement. Recall that solder will follow the iron because of its high temperature. Solder removal is the principle use of desoldering braid.

Solder wick tinning can be performed anywhere you have a soldering iron. Load some solder into the wick. Melt some new solder and let it flow into the braid. Don't use wick you worked with in cleaning circuits of old solder. The solder in this wick will have been heated and melted three times by the time you apply it to the board for tinning. Solder does degrade after two uses. Apply flux to the area you are going to tin. Use of a *paste flux* and applied sparingly is highly recommended. Liquid fluxes flow too

Figure 5-19
Solder wick is an effective way to tin a prototype PCB .

much and applying the very small amounts needed is troublesome. Place the soldering iron on top of the solder wick and drag the wick down the length of the pads or traces. Move the iron and the wick together, applying only enough pressure to the iron to keep the wick sandwiched between iron and board. The result will be a thin layer of solder along the formerly bare copper. Should you heat the tracks and pads excessively, or place too much pressure on the wick, you may lift the copper. Remember, eighty percent (80%) of the bond strength of the adhesive bonding board and copper can be lost during heating. Also, some surface mount component pads are so small that they can easily be lifted. **Figure 5-19** shows a trace that has already been tinned and an adjacent trace that is getting tinned.

Drilling of Through-Holes

Any board that has through-holes must have these holes drilled into it. If you are fortunate enough to have a 3-axis milling machine, the mill itself does all of this work and does it with great precision. However, one-off boards, or very small production runs, may be drilled entirely by hand.

A limited production run may require a technician to drill every hole using a micro drill press. The accuracy of the drill is paramount. A drill press that has run out (the tip of the bit moves when it contacts the work piece) is not suitable. Most holes that must be drilled for dual-inline-package ICs (DIPs)

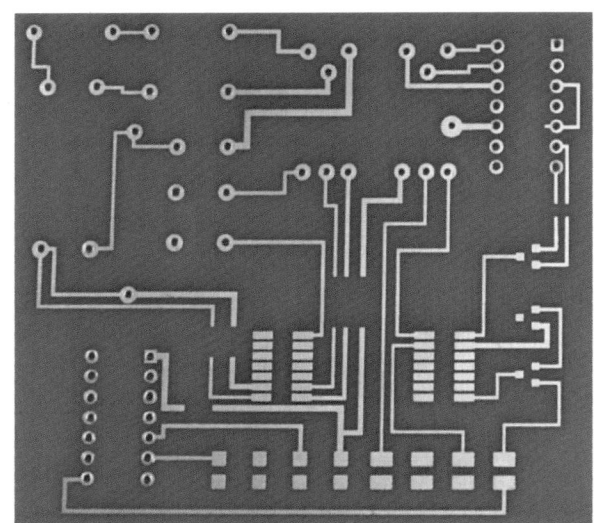

Figure 5-20
This is a new board which has just been drilled. A #67 bit was used.

are 0.032" (0.8mm) in diameter, less than 1/32" (see **Figure 5-21** for an example of an IC). The pad that this hole must be drilled through is also small. A misalignment of 1/32" (0.8mm) during the drilling process will ruin the pad, necessitating replacement. Additionally, the center-to-center distance between each leg of the IC is just 1/10" (2.5mm)! Visually aligning the bit and pad is still the most cost effective means of getting the job done. Some technicians like to mark straight lines through the pads on the board with a pencil or non-permanent marker to aid alignment of the drill. This can be done but be sure you get every bit of graphite off the board when done (see **Figure 5-22**). Graphite is good conductor and can cause problems if left between traces or pads.

If the holes to be drilled are of different sizes, you should be given a **Drill Schedule**. A drill schedule is usually an image of the board (like **Figure 5-13**) and will indicate which bits are to be used for each hole on the board. Two quick warnings about bits. All bits used on PCBs must be **carbide**. These drill bits are extremely small and fragile. Carbide tips also retain their edge over long use. Carbide bits are extremely delicate. Try to clear residual fiberglass from a bit with your fingers and you could easily snap it. Using a substandard bit may cause the inside edges of the drilled pads to lift because of the pressure and temperature generated. This is seen in **Figure 5-23**. **Table 5-2** lists some commonly used drill bit sizes in millimeters, inches, and their numerical equivalents.

Figure 5-21
A **DIP** (dual-in-line package). The scale on the ruler is 1/100" per minor division.

Figure 5-23
This is the result of using dull drill bits. Notice how the edges, where the drill cut, have flared up.

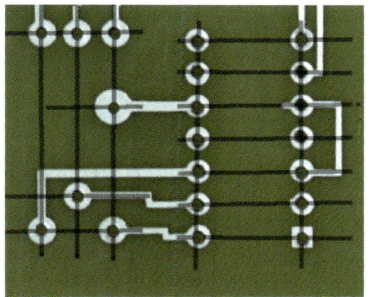

Figure 5-22
A non-permanent marker was used to draw lines to make drill alignment more accurate. This is fin e as long as the board is thoroughly cleaned afterwards.

Drill Bit Table		
.8128mm	.032"	#67
.8382mm	.033"	#66
.889mm	.035"	#65
.9144mm	.036"	#64
.9398mm	.037"	#63
.9652mm.	.038"	#62
.9906mm	.039"	#61
1.106mm	.04"	#60
1.0414mm	.041"	#59
1.0668mm	.042"	#58
1.0922mm	.043"	#56
Table 5-2		

Chapter 6
Installation and Removal of
Through-Hole Components

Once the circuit board is constructed, mounting components to the board becomes necessary. If repair is needed, removal and re-installation of parts is required. Installation and removal are straightforward processes if you are aware of the variables involved. These include the components mass as well as the type of pad the part is mounted on; single sided unsupported through holes, plated through holes, or through holes with hidden and blind vias. This chapter will discuss all of these connections and their effects on the soldering process.

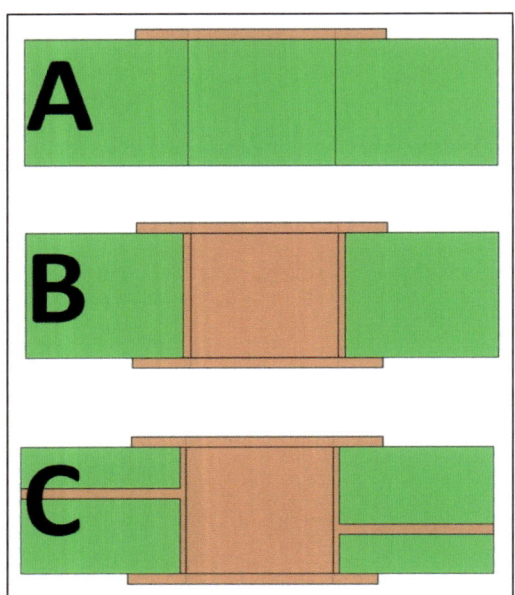

Figure 6-1
Three different board configurations; (A) single sided, (B) double-sided, and (C) multi-layer. Each requires a different soldering/desoldering technique.

What Does the Board Tell You?

When installing or removing components on a printed circuit board, you must know what kind of board you are dealing with. Is the board single layer unsupported, plated through-hole, or plated through–hole with hidden or blind vias. Each type must be approached differently. **Figure 6-1** is a drawing of the three types of boards mentioned. The non-supported through-hole (**NTH**, item A) will require the least amount of time to install and remove because of the relatively low thermal mass it presents. As you progress to plated through-holes (**PTH**, item B), the mass increases requiring an adjustment in heating. However, all of the through-holes series should take no more than two to four seconds of dwell time to install or remove components. Much longer than this and you risk lifting the single sided pad (remember, it is held into place with an adhesive that weakens considerably when heated and looses all adhesion when heated and placed under pressure). You may also ***measle*** the board. Measling occurs when the iron has heated the substrate to such a

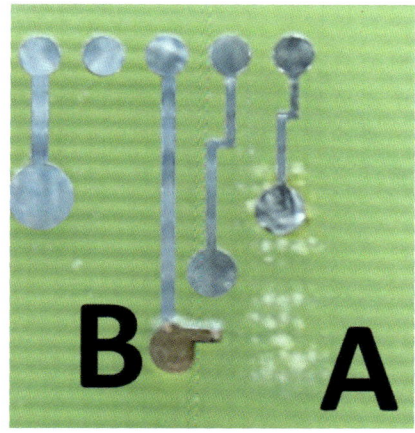

Figure 6-2
All the white spots are damaged and will need repair. The horizontal lines are the weave of the fiberglass in the board.

degree that the resin in the fiberglass **overcures** (the resin becomes powdery and separates the weave of the fiberglass). This creates weak spots in the board. An example of measles resulting from excess heat can be seen in **Figure 6-2**. The white spots around point A in the image are measles. These can be difficult to see. To make them more visible, apply a small amount of the cleaning solvent you are using. This will make them stand out against the background. Point B shows a pad that was completely lifted and has been flipped over. The copper visible to the right of point B is actually the bottom of the destroyed pad. If you see this kind of damage, the board must be repaired. The repair is time consuming (which makes it costly) and delicate, so it is best to avoid causing such damage in the first place.

Gravity, Surface Tension, and Capillary Action

Gravity, surface tension, and capillary action are important considerations in soldering and repairing different board types. The affects of gravity, surface tension and capillary action vary with the size of the solder connection. However, on the small scales involved in electronics soldering, capillary action has the greatest affect and gravity the least. A concave fillet can be made while soldering upside down because of surface tensions dominance over gravitic effects. Capillary action even occurs if you are soldering on a board at arm's length above you. How all these forces affect your work is discussed in the following paragraphs.

Gravity – This is the smallest force acting on the board during hand soldering. Wave soldering devices, however, must be

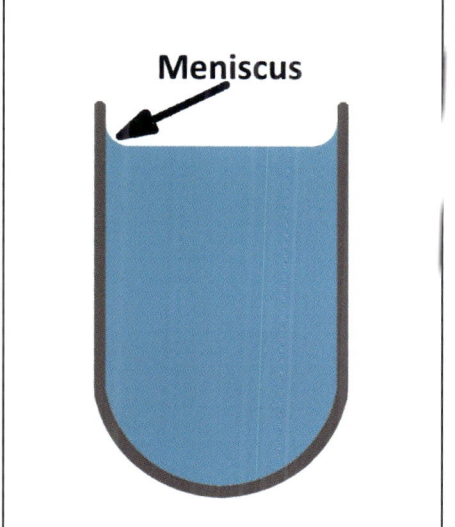

Figure 6-3
Example of how a meniscus appears in a tube filled with water. Molten solder behaves the same way on metal.

set up to compensate for gravity. These systems are used in large scale production and work as the name implies; they send a wave of solder down a channel or through a slot. The wave is just high enough to reach the pads and solder the connections. Since the solder side of the board is face down, gravity can cause the solder joints to pull down slightly. This may leave solder connections that lack the

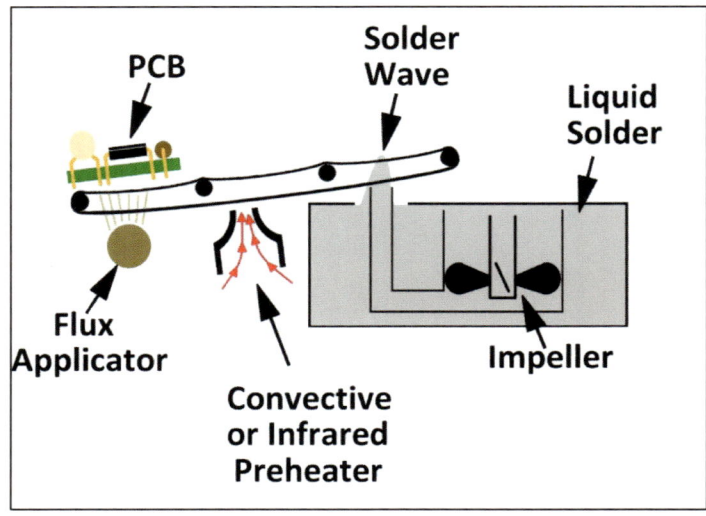

Figure 6-4
A simple model of wave soldering.

preferred concave fillets. Proper flux, applying the correct amount of solder at proper temperatures compensate for gravities influence.

Surface Tension –

Surface tension plays a considerable role in soldering all types of connections. Surface tension is the attraction a material, solder in our case, has for the material it is placed in contact with – the copper traces, pads, and component leads. You can see the same attraction between water and glass. The water forms a concave meniscus where its surface contacts the glass. This is the same shape as our concave solder fillet (see **Figure 6-3**).

Capillary Action – Capillary action is closely related to surface tension. In capillary action, a liquid is drawn up a tube because of surface tension. The narrower the tube, the greater distance the liquid is drawn into the tube. This increased attraction and distance is capillary movement. Think about the spaces in a stranded wire. They are essentially very narrow tubes. Due to their narrow width, molten solder can be drawn into the wire for some distance. This is the cause of wicking and the reason solder wick works.

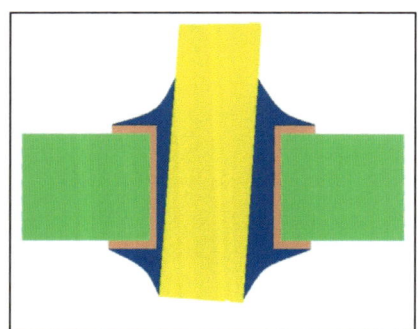

Figure 6-5
A soldered PTH has surface tension and quite a bit of capillary action.

Surface tension and capillary action are the primary elements to consider in soldering and desoldering plated-through holes. The yellow area in **Figure 6-5** represents a lead going into a PTH. Notice that the lead is isn't exactly vertical. This is typical of most leads, very few are absolutely perpendicular to the board. This causes the capillary action in the PTH to vary along the component's length (the "tube" is narrower on one side). It also causes variations in surface tension. Surface tension and capillary action will be greatest where the distance between the lead and PTH walls is smallest.

94

Figure 6-6
Single sided boards have surface tension but little capillary action.

Single sided boards will be affected most by the surface tension of the solder. The concave fillet desired on almost all soldering is a result of proper heating of the component and pad, application of the correct amount of solder, good wetting, and proper movement of the iron when finishing the solder joint. Performing any of these points incorrectly easily overwhelms the weak force of surface tension. Again, the fillet is an excellent indicator of proper technique. If you look carefully at **Figure 6-6**, you can see a small amount of solder has flown into the hole. This was pulled in principally by the surface tension of the component lead, less so by gravity.

Before Component Installation

Once you know the type of circuit board you are going to work on, you must prep the component. This includes applying the proper bend radius to the leads, forming mechanical lead terminations as required, applying any necessary stress reliefs, and cleaning. Only then can the component be correctly mounted and soldered to the board.

Lead Bending

Forming a lead with too severe a bend can be a serious problem. The arrow of **Figure 6-7** shows a set of leads that were bent closer to the body of the component than is acceptable. The point where the lead enters the component body is mechanically weak. Also, the point at which lead and component join is an environmental seal. Damage this junction and the life of the part can be shortened. Electrical characteristics may also be altered by such damage. **Figure 6-8** shows several components and the points at which bend *cannot* be made. Component A, a capacitor, has a meniscus that can be damaged by lead manipulation. Component B is also a capacitor but this piece has a small weld. Lead bends cannot be made too close to

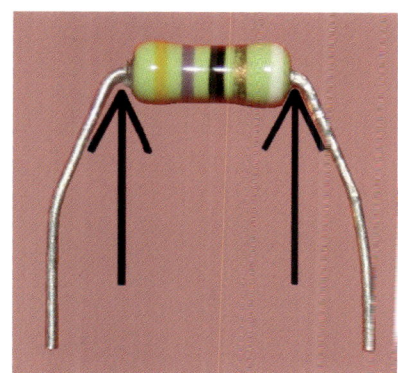

Figure 6-7

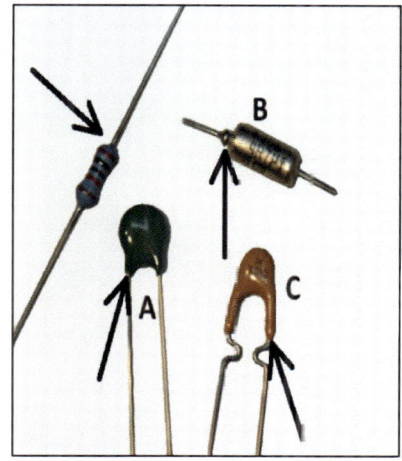

Figure 6-8

this point either. The meniscus of part C should not be damaged. This component has lead bends that were added at the factory. **Table 6-1** gives the standards for the minimum bend radius and distance of the bend from the component body as they correlate to the American Wire Gauge values. This table also determines how close a bend can be to the meniscus/weld/body of the component. For example, a component with 20 AWG leads should have the leads bent at least 1.5X the lead width from the body and the bend should have a radius of 1.5X of the lead diameter.

Lead Diameter	Bend Radius and Distance
Smaller than 20 AWG – 0.031" (0.8mm)	1 diameter
16 to 20 AWG – 0.031" to 0.047" (0.8mm – 1.2mm)	1 1/2 diameters
16 AWG or larger wire – 0.047" (1.2mm)	2 diameters

Table 6-1

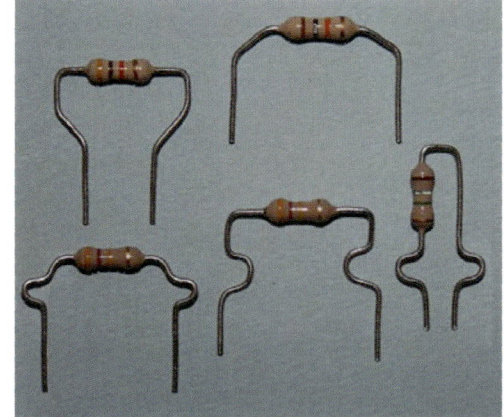

Lead Forming – A lead form is a bend in a component's leads that provides mechanical support to that component. Lead forming is required only on boards with unsupported through-holes, usually single sided boards, and only when the component is not directly on the board. A lead form is not required if the part is less than 0.028" (0.7mm) off the board (about the thickness of a 22 AWG wire). Lead forming is performed with the same constraints as lead bending; any bend performed has to meet requirements of **Table 6-1**. A few lead forms are shown in **Figure 6-9**. This photo is not all inclusive, many varieties exist. All shown can be made with a set of round nose pliers or, given the budget, pliers that have jaws to form the leads all at once. Lead forming is also performed with automated devices. **Figure 6-10** shows some lead forms in an unsupported single sided board. When making lead forms or bends, damage to the wires must be avoided. However, as long as damage does not exceed 10% of the width or thickness of the conductor, the work is generally acceptable.

Figure 6-9
Lead forms must be used on unsupported through-holes where the component is not in direct contact with the board.

Figure 6-10

Lead Forming Tools – Using round or needle nose pliers on a continuous basis for lead forming becomes tedious very quickly. Several companies manufacture relatively inexpensive tools to make repetitive lead forming less tiresome. The simplest of these is the "pine tree" (gray tool in **Figure 6-11**). It relies on the "calibrated eyeball" to align the tools sides with the holes into which the component will fit and take note the numbered slot that aligns with those holes. You then insert the component into that slot number and bend the leads. The tool is suitable for axial lead components only.

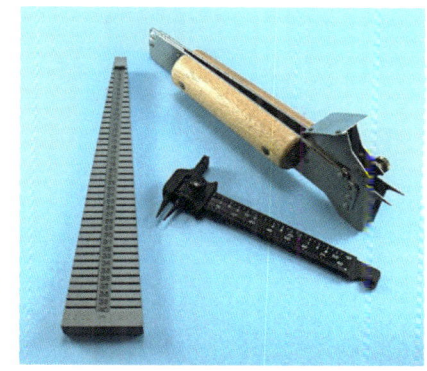

Figure 6-11
A few of the lead forming tools available.

The lead bending tool with the wooden handle is somewhat improved. Place a point of the tool into one of the leads holes and adjust the wheel until the second point fits into the other hole. Place the component into the built in holder and bend appropriately.

The last lead forming tool in the group offers the greatest functionality. This tool, from PACE Inc., allows you to not only measure distance between the through-holes but also add stress reliefs (stress reliefs will be discussed later). The tool can also be used on axial and radial devices.

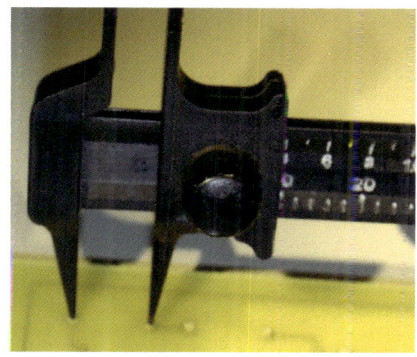

Figure 6-12
Take measurements from the board's component side.

Mounting an Unsupported Through-Hole Component

Through-hole components that are not set more than 0.028" from the board do not need a lead form. They will require lead bends. Begin by using one of the lead bending tools or some pliers (without serrations). We will use the Conform Tool by PACE® to measure the distance between the holes into which the part will be placed. Always take these measurements from the component side of the board (**Figure 6-12**). Loosen the wheel and fit one point

Figure 6-13

97

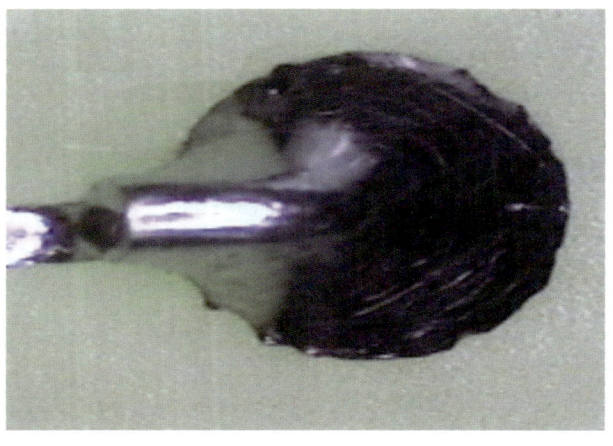

in each hole of the lead. Once the distance is measured, lock the wheel. Insert the component into the tool and bend the leads (**Figure 6-13**). Always support the body of the component to prevent damage at the meniscus or weld and use a non-destructive tool when bending wires! Once bent, the leads should slide into the holes easily. Ideally, you should be able to tap the board with your finger and the component should bounce in its holes.

The leads will need to be trimmed after insertion. Because this is an unsupported through-hole, the leads must also be clinched. Clinch the leads before trimming them. Clinching refers to bending the leads over on the solder side of the board. A lead can be semi-clinched – bent to a 45° angle as in **Figure 6-15** or fully clinched – bent to a 90° angle as in **Figure 6-18**. Whichever clinch is used, it must have sufficient clearance from adjacent leads, tracks, pads, etc.

Unsupported through-holes must have a semi-clinch as a minimum. Use a plastic or wood tool, such as an orangewood stick (**Figure 6-14**) to bend the lead. This prevents damage to the conductor. When the lead has been clinched, trim it so the end of the lead doesn't extend beyond its pad. If possible, clinches should be bent in the same direction as the track they attach to. Once everything has been appropriately cleaned, the leads can be soldered. As always, use flux.

The correct soldering iron tip should be no wider than the pad on which it will be used. Too wide and you can damage the substrate. Keep in mind that you should have the correct size and type of solder on hand. Most pads for through-hole components will be buried in excess solder if you use one 0.05" (12.5mm) wide. 0.031" (0.8mm) is a good starting size and you can adjust from there. Place the iron, with a small bit of solder on the tip (to form a heat bridge) at the point of greatest mass (**Figure 6-15**). This will be the junction between the pad and the lead. Once the flux has been activated, apply enough solder to the pad to achieve good wetting. When finished, wipe the soldering tip *up* along the lead to help form a concave fillet. This also insures the end of the lead is covered in solder. It is desirable to have the tip of the lead covered in solder but it is not a requirement (unless specified by the organizations guidelines). Trimming the lead after soldering leaves copper exposed to air which may lead to deterioration. *Conformal Coatings*, which will be discussed later, can help prevent this from occurring. Thoroughly clean the connection when finished.

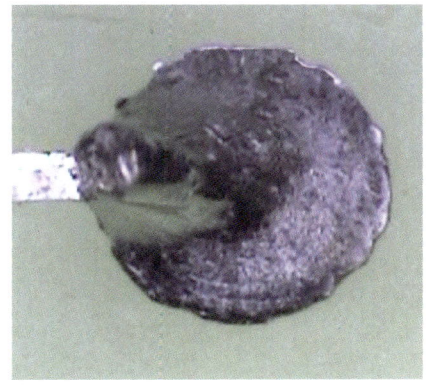

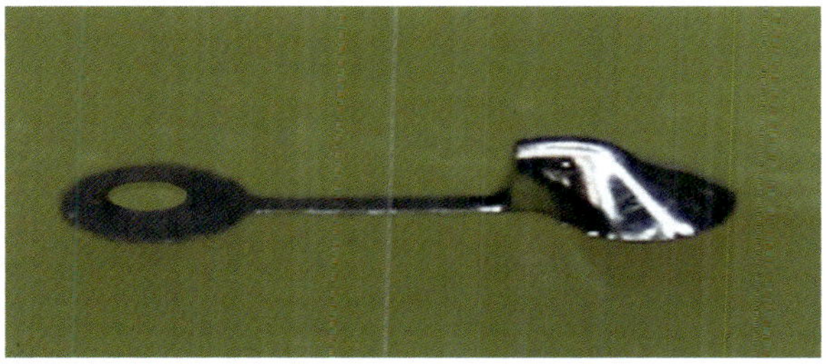

Figure 6-18
An ideal solder connection with leaded solder will have this appearance.

Figure 6-17
A lead done with lead-free solder.

A good solder connection should look like the images of **Figure 6-16**, **Figure 6-17**, and **Figure 6-18**. **Figure 6-16** is a connection done with 63/37 solder. Notice that it is smooth and shiny. **Figure 6-17** was done with a lead free solder. Lead free solders tend to have this mottled appearance so this is an acceptable solder joint. Both types of solder should have a concave fillet. Notice that fillets are present at the heal *and* toe of the solder joint in **Figure 6-18**. The solder on the toe (left side of soldered pad) should not run down the trace.

Figure 6-19
An extreme example of a cold solder joint.

Mistakes in Through-Hole Soldering

Problems with soldering, whether on wire connections or through-hole components, do not change much. Both can have, disturbed joints, flux pitting, poor wetting, cold solder connections, etc.

Figure 6-19 is an example of a cold solder connection. The connection will look grainy and may exhibit voids. Insufficient heating can also lead to flux pitting as shown in the first circle of **Figure 6-20**. Flux pitting is the result of too little heat during the soldering process. This causes unburned flux to remain on the surface. This will leave a depression in the solder after the joint is cleaned.

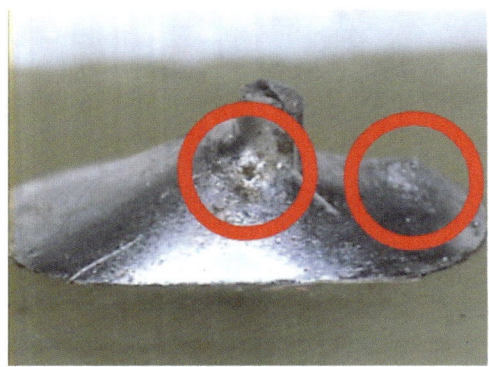

Figure 6-20
Pitting caused by flux and a mound of solder created because the iron was drawn off the work improperly.

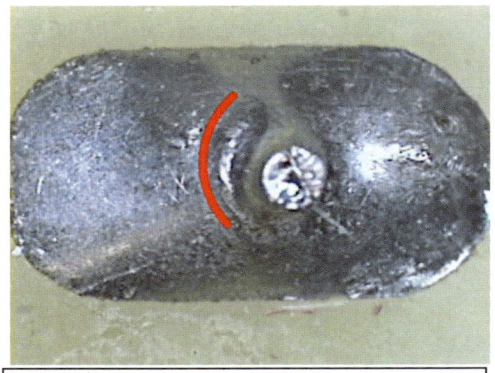

Figure 6-21
The connection moved during the plastic state and created this disturbed joint.

Figure 6-22
Damage to the base metal. This component will have to be replaced.

The second circle indicates the iron was pulled away from the work incorrectly. Remember that solder follows heat. Drawing the iron away from the side (versus wiping up along the lead) can cause this bulge. **Figure 6-21** is the same lead as **Figure 6-20** viewed from the top. Notice the area to the right of the red arc. This line indicates that there was movement of the lead during the solder's plastic state. Such deformations are unacceptable. The last example shows damaged base metal (copper lead of **Figure 6-22**). Base metal damage to the lead, pads, traces, etc is not acceptable.

Plated Through-Holes

Installation of components in plated-through holes (PTHs) is slightly more involved than single-sided board work. First of all, the component's meniscus or weld cannot rest on the pad. The fillet that will be formed on both sides of the board may contact the component body. This is not acceptable. It becomes necessary to mount the component off the board a minimum of 0.059" (about $1/17^{th}$ of an inch or 1.5mm). Because PTHs are supported through holes , lead forms are not required.

As always, clean the pad and lead terminals well. Use a good flux to aid in oxide removal and efficient pad wetting. Place a small amount of solder on the tip of the iron to form a heat bridge and

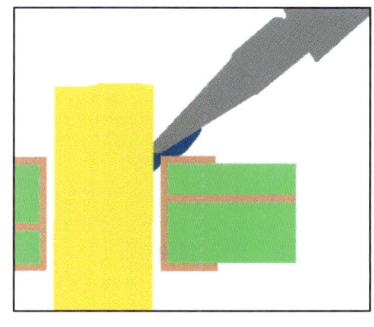

apply the iron to the point of greatest thermal mass; the junction of the component lead and pad (**Figure 6-23**). Add solder to the heated pad once the flux has activated. You will see the formation of a solder fillet (**Figure 6-24**) and then it will suddenly drop (**Figure 6-25**). Capillary action and surface tension have pulled the solder through the plated hole. A fillet will now form on the component side of the board. Add a little more solder to create a fillet on the solder side and then wipe upwards along the component lead. The solder joint should look like **Figure 6-26**. Wiping up along the lead draws the solder to the top of the lead and assures that It is tinned.

Figure 6-23
This is illustrates the proper placement of the soldering iron tip.

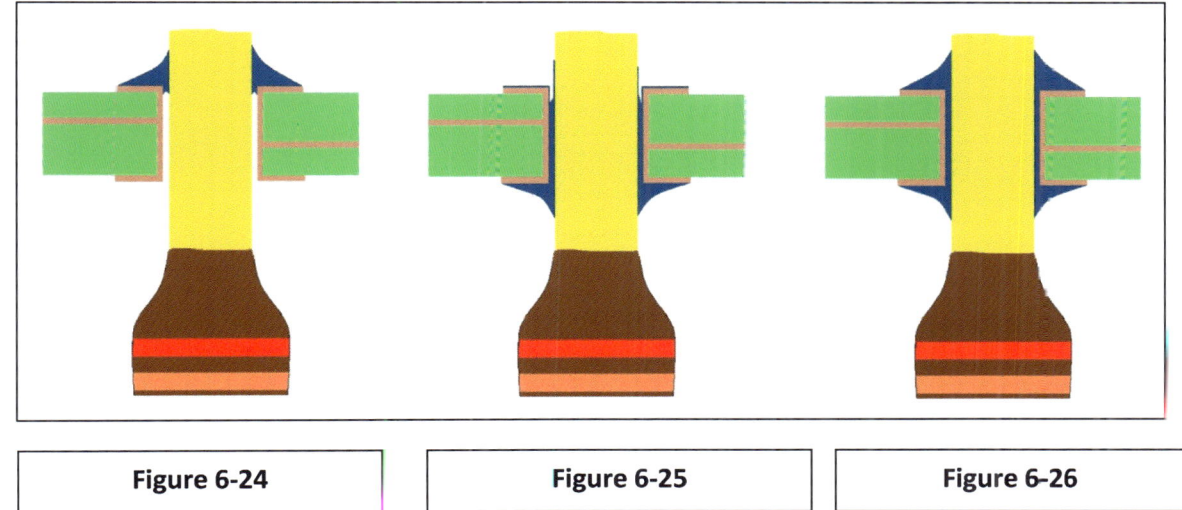

| **Figure 6-24** | **Figure 6-25** | **Figure 6-26** |

This standard of soldering exceeds the Class 3 standard established by IPC J-STD-0010. The requirements of the standard are: 1) the vertical fill of the plated through-hole is required to be 75% and 2) there must be 330° of wetting on the solder source side, 270° on the destination (component) side.

Stress Reliefs

Stress reliefs are bends on component leads designed to minimize physical stresses (vibration, inertia, etc) on that component. They should only be used with PTH boards as the PTH supplies proper lead support. The stress relieved component cannot rest on the board. Placing the part on the board defeats the purpose of stress isolation. Stress reliefs have as many varieties as lead forms. You will have to check your organizational directives for the type desired. Rework and repair requires that a new component have the stress relief inserted by the technician. Not many dedicated hand tools are still manufactured for this sole purpose anymore. However, round nose and duckbill pliers can always be used. The tool used here a Comform I from PACE®. Begin by measuring the distance between the PTHs. Note the measurement on the upper scale (**Figure 6-27**). Subtract 2 minor divisions (using the upper scale) from this reading (**Figure 6-28**). Place the component in the notches on the tools handle and bend the leads

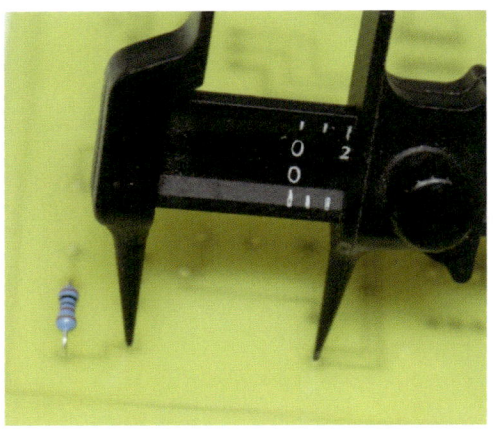

Figure 6-27
Measure distance between the holes – use the upper scale.

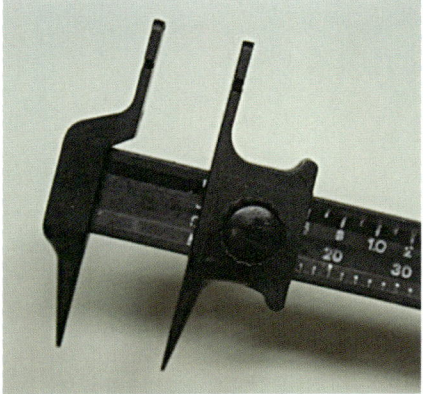

Figure 6-28
Subtract two minor divisions from the reading noted in **Figure 6-27.**

down so they are perpendicular to the component body (**Figure 6-29**). Next, take the part and place the leads in the notches at the end of the Comform's handle as shown in **Figure 6-30** and now bend the lead around the notch (indicated by the red arrow). Repeat this procedure for the other side. The lead bends should be done with a non-destructive tools, such as the orangewood stick shown in **Figure 6-31**. Each bend adds 1 minor division back to the total length of the component lead. This restores the part to the original width making insertion into the PTH easy. The part should fall into

the holes with little friction and the leads must be as close to 90° from the plane of the board as possible. Finally, when the part is soldered in, fillets should be seen on both sides of the board (**Figure 6-32**). All the standards for a good solder joint apply.

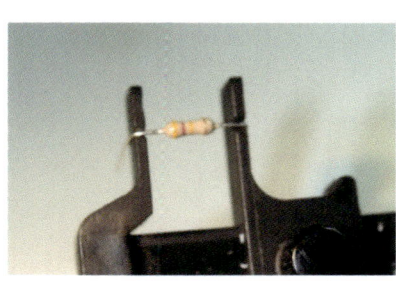

Figure 6-29
Place the component in slots and bend the leads.

Figure 6-30
Use the notches at the base of the tools handle to insert the stress relief.

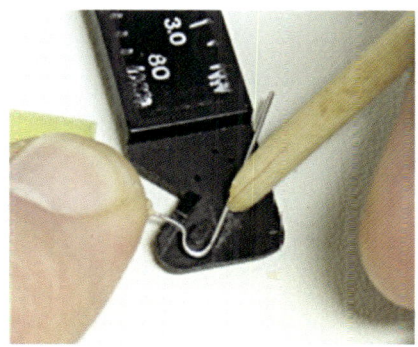

Figure 6-31
Adding the stress reliefs re-inserts the two divisions subtracted from the scale earlier. Use a non-destructive tool to form the leads.

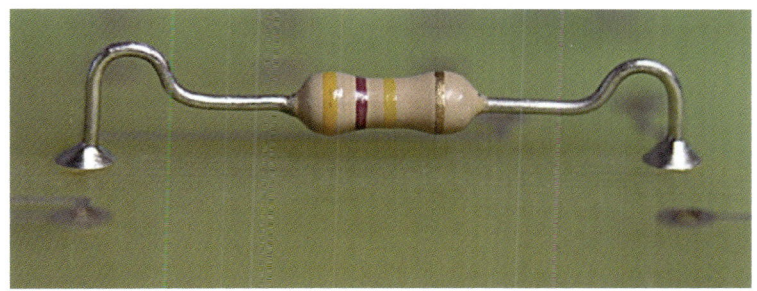

Figure 6-32
A resistor with stress relief on a PTH board.

Dual-In-Line Packages (DIPS)

Installing a Dual-Inline-Package (DIP) integrated circuit (IC) is just like soldering an unsupported or plated through-hole component. The exception is the number of leads that must be soldered. DIPs can have as few as 4 pins to more than 40. The only new guideline is that you must not solder adjacent pins. Look at **Figure 6-33**. The pins labeled 1 and 2 should be soldered first. If the pins are in unsupported through-holes they should be semi or fully clinched. Once these pins are soldered, skip around, go to the pins marked 3, then 4, wherever you wish but _don't solder pins next to each other_. Doing so can cause heat buildup in a small area leading to damage of the component.

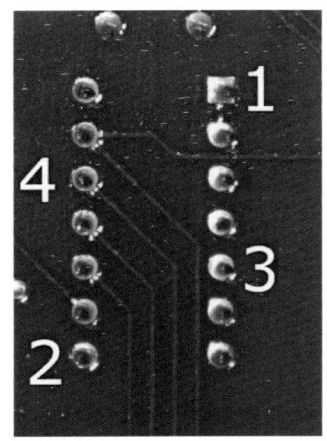

Figure 6-33
Skip around the pins to avoid building up heat in one spot.

You must know what kind of IC you are dealing with before installation. The majority of ICs are now Complementary Metal Oxide Semiconductor devices, commonly known as CMOS. These devices are electrostatic sensitive. The small charge people normally carry on them is sufficient to damage (destroy) the component. Make sure to use ESD precautions (see **Appendix A**). Never touch the component when you are not ESD safe and leave it in its protective package until you are.

Pin 1 in the figure is also pin 1 of the device. Pins 2 would be directly below pin1. The pins on an IC (when viewed from the pin side of the device) are counted in a clockwise direction.

Heat Sinks

Many technicians use heat sinks when soldering. These are clips that attach to the lead you will work on to dissipate heat. Heat sinks are not suited for use on ICs. The clips are too large. Using a heat sink on axial leads, radial leads, or for tinning is unnecessary for a well trained technician. They can act as a crutch and lead to sloppy soldering techniques. However, if your organization requires there use, follow their rules.

Through-Hole Component Removal

Conformal Coatings

Conformal coatings are materials placed on the surface of PC boards to protect the board against environmental contaminants. On occasion, the conformal coating is used to increase the dielectric constant of the board, i.e. to increase its resistance to undesirable electric fields. To remove a component or perform a board repair, the

protective coating must be correctly removed. To do this you must know what the coating is.

Conformal materials come in a variety of types. These include, epoxies, acrylic lacquers, silicon, polyurethanes, Parylene (also known as Paraxylylene), and innumerable proprietary formulas. The methods of application are almost as numerous. They can be sprayed, dipped, vacuum deposited, brushed, and robotically spot applied.

The material characteristics and application methods are good indicators of what the coating is. For example, if you find a thick material that is shiny and *cannot* be dented with your fingernail, you can reasonably assume that the conformal material is epoxy. If the material is thick, shiny, and *can* be dented, it is most likely polyurethane. By using the information in **Table 6-2**, you can narrow the possible conformal materials down. This is hardly a definitive listing. Nothing beats knowing exactly what was used, especially when proprietary materials were applied. Proprietary conformal coatings require proprietary solvents for removal.

	Thick or Thin Coating	React to Alcohol	Reacts to Heat	Forms a Powder When Heated	Feels Rubbery or Soft	Shiny or Dull Appearance	Can Be Dented or Scratched
Acrylic Lacquer	Thin	Yes	Yes	No	No	Shiny	No
Epoxy	Thick	No	Yes	Yes	No	Shiny	No
Parylene	Thin	No	No	No	No	Dull	Yes
Polyurethane	Thick	No	Yes	No	Yes	Shiny	Yes
RTV Silicones	Thick	No	No	No	Yes	Dull	Yes
Silicon Resin	Thin	No	No	No	Yes	Shiny	Yes

Table 6-2

Removal Methods

Once you identify a material, you must know how to remove it. The image in **Figure 6-34** shows an epoxy covering a component (the opposite side of the board is also protected by epoxy). You can see that the material is thick and shiny. This narrows the material down to polyurethane or epoxy. If it has a rubbery feel, it

Figure 6-34

is polyurethane. In our case, it is hard identifying the substance as a possible epoxy. If it reacts to heat and forms a powder, it is epoxy. You can use a soldering iron with an old tip and touch the epoxy to see if it creates a powder. You can even use the tip to overcure and remove the epoxy (an epoxy that has dried and hardened is said to have *Cured*, **Overcuring** is heating an epoxy or resin beyond a set temperature which results in the physical breakdown of the substance). Do not apply too much heat. Epoxy that turns brown has been burned. Using a soldering iron is not the recommended method for coating removal. The tip is too hot and too hard and you may seriously damage the board.

So, we have an epoxy. How do we remove it? Proper tools are a plus. In **Figure 6-35** you see two tools. The top device forces heated air out of the tip. Both the heat and air volume are adjustable on most machines. A switch in the handle activates the device. The air outlet can be changed to adjust the pattern of the air flow. This hot jet tool is used for the finer work. The smaller tool has a tip whose temperature is also adjustable. This tool, called a *Lap Flow* by the manufacturer (Pace Inc.) is used to do the coarse work of removal. By applying this tip to the work, the epoxy overcures and can be pushed away. This "shoveling" must be done gently, about the amount of pressure that it might take to break a toothpick.

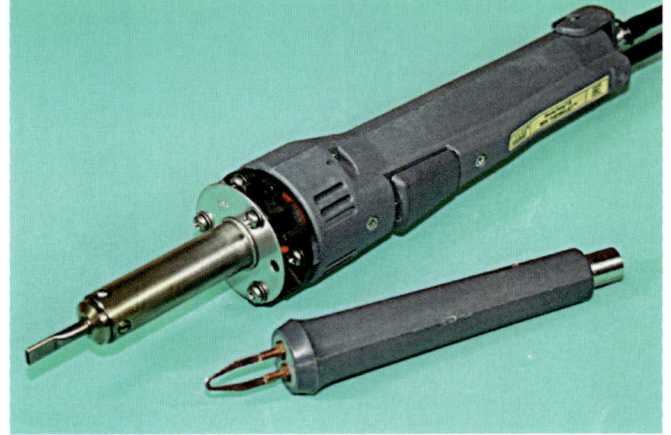

Figure 6-35

Figure 6-36
Using a lap flow tool to overcure epoxy.

Figure 6-36 shows the beginning of the removal process. The inset shows how epoxy coatings turn to powder when heated. Be sure not to touch the board, leads, or component with the tool. Doing so will lead to damage that will also need repair.

At some point, you will have to transition to the hot air tool. This point comes when using the Lap Flo anymore will cause damage to the work piece. The hot air tool and a wooden push tool (cuticle sticks work well here) will be used to remove the remainder. Begin by setting the hot air jets temperature. A rule of thumb is that the air from the tool should turn a paper towel or piece of facial tissue brown after two or three seconds of exposure from a distance of about 1/2". Start at 350° (176°C) and adjust from there. When the tool is ready, apply the hot air to the area

Figure 6-37
A hot air stream does the more delicate work.

being worked on. After two or three seconds, use the stick to remove any epoxy that has overcured (**Figure 6-37**). You can stop applying heat while using the dowel. If you cannot remove any of the conformal coating, increase the duration of heating. Heat, remove material, and repeat as necessary to clear both sides of the component.

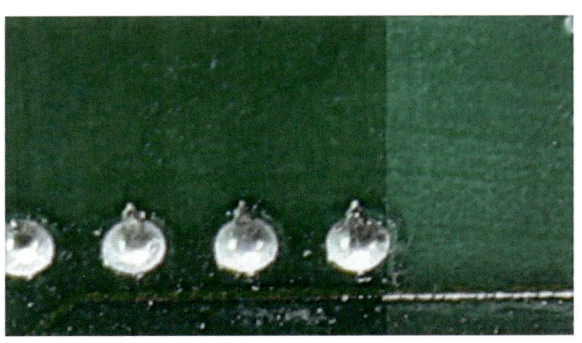

Figure 6-38

In **Figure 6-38**. Only a portion of the board has been covered by a conformal coating. The material is shiny and thin. This narrows the material down to an acrylic lacquer or silicon resin. Application of some alcohol turned the connection milky. Because it reacted to alcohol, it is an acrylic lacquer resin.

There are several ways to remove acrylics and silicon coatings. Abrasion is quite effective. Using an abrasive brush or bullet with a handheld rotary tool (Dremel® or a Microchine™ hand piece) is the safest and easiest method for most work. Set the speed of the hand piece low, 1000 RPM or thereabout. The first picture, **Figure 6-39** shows a brush being used. The brush is strong enough to remove the coating but, when applied with a light touch, will do no damage to the board or traces. You will probably want to tape around the area in which you will work. This defines the area and assures you only remove the necessary material.

Figure 6-39
Using a brush to remove an acrylic coating.

An abrasive bullet can also be used (**Figure 6-40**) but must be applied cautiously. It has considerable more grit than the brush and has little "give." Press too hard and you will remove traces and substrate. Always use the least destructive method for any repairs!

Sometimes solvents are the only choice available for coating removal. If this is the case, be absolutely certain that you are using the right one and tape off your work area. **Figure 6-41** shows the result of incorrect solvent selection. It is the same board we have been working with. Recall that it is an acrylic based coating. It may react with alcohol but that

Figure 6-40
A bullet can also be used to remove the coating but be careful; it can remove the board too.

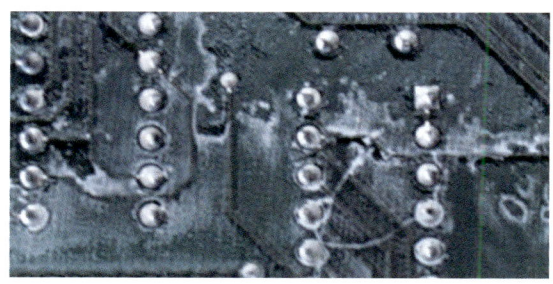

Figure 6-41
The result of using the wrong solvent. All if this material will have to be removed.

doesn't mean alcohol is the preferred solvent. This happens to be a proprietary acrylic coating and the manufacturer recommends another solvent for proper removal. Solvents also have a tendency to go where they are not wanted. This runoff does an excellent job of removing coatings where you do not want them removed. Any areas that the runoff comes in contact with will need to be stripped and recoated. Solvents are most useful if the entire board has to be cleared of a protective coating. If you are working with an epoxy conformal coating, don't use a solvent. Remember, the board is epoxy and fiberglass. Use an epoxy solvent and your board will be a mush of fiberglass and copper strands!

Methods of Component Removal

There are three common methods for removing through-hole components: solder wick, air impulse, and constant airflow. Each of the three approaches has benefits and shortcomings. These methods, plus their applications, will be discussed in the following paragraphs.

Solder Wick   Solder wick (also called solder braid and desoldering braid) is a woven and flattened collection of fine copper wires (Figure **6-42**). It comes in many lengths but spools of 5' and 25' (1.5m to 9m) are most common. It can be used directly off the roll but is much more effective if a bit of paste flux is added. The paste can be applied precisely, without the worry of having liquid flux running across the board, and seems to leave the board a little cleaner. The wick works just as its name implies. The strands are very close together in the braid. This causes the braid to have a strong capillary action. To use it, take a small amount of paste flux and rub it into the last two to three inches of wick. Place the wick (hold it with a pair of needle nose pliers – the wick gets quite hot) on the point to be desoldered and put the soldering iron on the top of the wick.

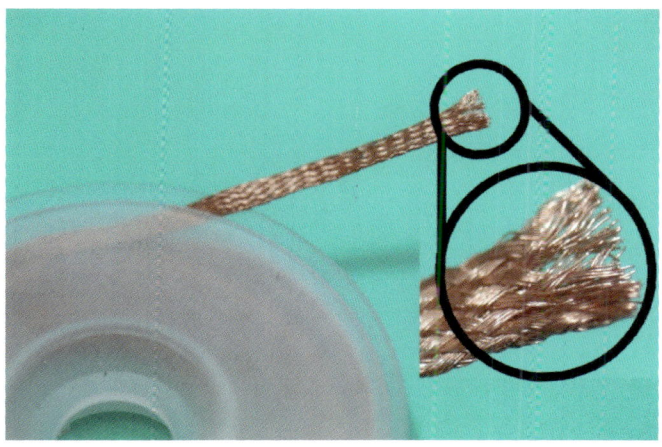

Figure 6-42

After a few seconds, the solder will begin to flow into the wick. You can see this procedure in **Figure 6-43**). If the soldered point has a large amount of solder, the wick will have to be moved to a fresh section along its length. Once the wick saturates, it loses its effectiveness.

Figure 6-43

Solder braid has one significant shortcoming. It works by way of capillary action. A plated through-hole also exhibits capillary action (as well as surface tension) and is affected by gravity. Solder wick is unable to overcome all of these bonds and has a tendency to leave solder in the PTH (**Figure 6-44**). This small amount of remaining solder is called a *Sweat Joint*. A strong mechanical connection remains between the lead and the wall because of this residual solder. Forcing the lead out will damage the wall of the PTH necessitating its replacement. For this reason, solder wick is best used on single sided boards and for trace cleanup.

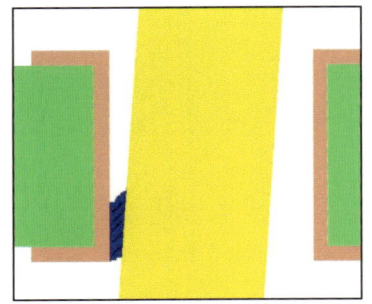

Figure 6-44
Solder wick is less effective with plated-through holes.

Air Impulse

Air impulse devices work by creating an impulse of low pressure (a vacuum). By squeezing and releasing a bulb or a charging and releasing a trigger, a vacuum pulse is created which draws in molten solder. A bulb and solder extractor (also called a solder sucker, solder pull-it, or vacuum extractor) are shown in **Figure 6-45**. The white tips of both are heat resistant and replaceable. Both also produce large static charges, up to 5000V so they should not be used on ESD sensitive equipment.

Both of the air impulse devices work in the same way. are used much the same. However, the bulb requires a little more dexterity because releasing the bulb to create the vacuum can cause the tip to move. The solder extractor has a trigger on its side and is easier to control. In both cases, apply the iron to the point to be desoldered and wait for complete melt. Be sure to have the tip of the desoldering tool

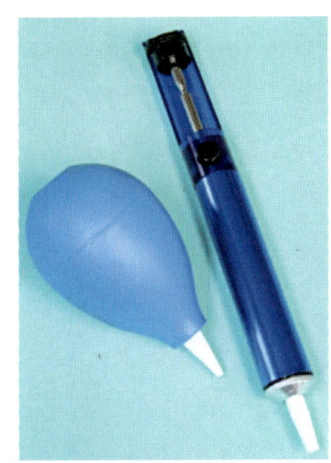

Figure 6-45
Two very common air impulse solder extractors.

Figure 6-46
Place a little solder on the irons tip to help transfer heat. Have the iron nearby to minimize the time the molten solder has to cool and the heating time.

Figure 6-47

close to the point to be desoldered. Once melt has occurred, release the bulb pressure or the solder suckers trigger and extraction should occur. Both tools work well on single sided boards and marginally well on PTHs. If complete solder removal does not occur, which is common in PTHs, add some solder to the joint and try extraction again. **Figure 6-46** shows a bulb being used while **Figure 6-47** shows a solder extractor. As you can see, both are used in much the same way.

Maintenance of these tools is relatively simple. As they are used, old solder builds up in the device. In the case of the bulb, just pull the tip out and tap to remove the debris. The extractor with the trigger will have to be disassembled. The majority of vacuum extractors can be disassembled the ways. The image of **Figure 6-48** shows the extractor that was used in the previous photos. To disassemble it, hold the top and the blue tube. Turn the tube (lower half) counter-clockwise. This removes the top (above the black ring) and allows you to pull the plunger assembly out. You can see the debris that was removed. The vacuum cylinder (blue tube) must be cleaned of residual solder; a paper towel will do the job. A small amount of mineral oil on the towel helps remove stubborn solder and improves the seal of the o-rings. Notice the two o-rings on the plunger. It is important that these be cleaned well. Use a paper towel and apply *mineral oil* (or whatever the manufacturer recommends) to the o-rings and reassemble the device. Charge and fire the tool two

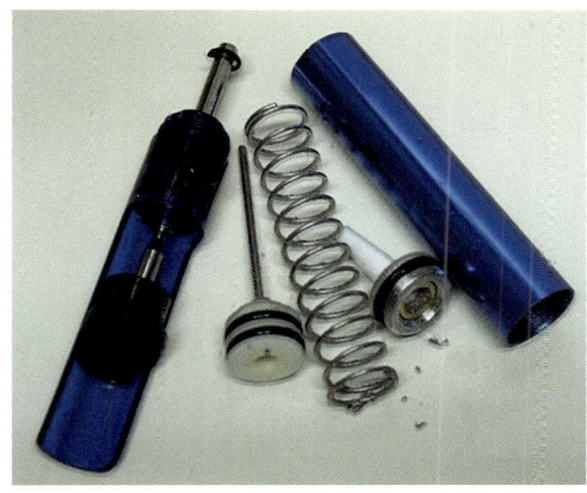

Figure 6-48

or three time and it should be restored to original condition.

Constant Vacuum

By far the best way to remove components is via a constant vacuum system. This approach uses a heated tip with a hole in the center. The hole provides an on demand vacuum and path to a collection chamber. This vacuum can be maintained as long as desired via separate pump housing. The housing also provides the power to heat the tip. Tips for the desoldering tool are also interchangeable. You will not want to use a tip larger than the pad you are desoldering. You also don't want a tip whose opening is too large. Chose a tip with; 1) an opening slightly larger than the *pin* to be desoldered and 2) with an overall width that just barely covers the pad. You can see two of these tips in **Figure 6-49** and the desoldering tool itself in **Figure 6-50**.

Constant vacuum systems are excellent for all uses. They de-solder PTHs very effectively and leave few sweat joints. The strength of the vacuum overcomes surface tension, capillary action, and gravity effectively (**Figure 6-51**). If a sweat joint is found, apply some solder to the connection and extract again.

The technique for using constant vacuum tools is simple. First, make sure that all inline filters are present and clean. Vacuum pumps do not like to ingest solder. Check that the vacuum system is functioning by placing the tip close to a paper towel or piece of tissue. When activated, the extractor should apply just enough vacuum to lift the tissue two inches or so before it falls away. Place the tool on the

Figure 6-49
Just two tip sizes available for a constant vacuum extraction tool.

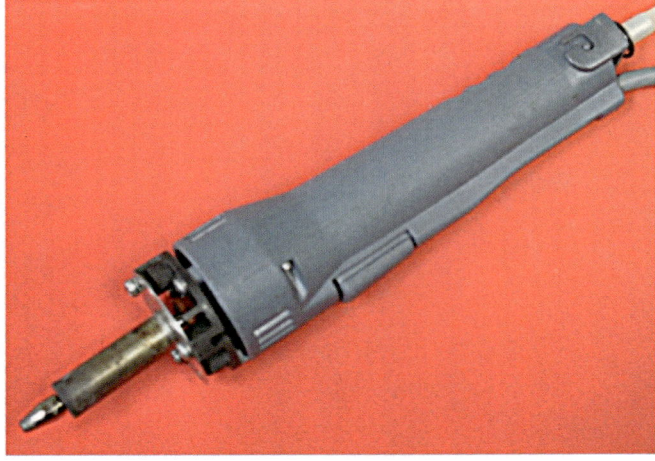

Figure 6-50
An example of one manufacturer's solder extractor.

lead as vertical to the board as possible and wait for solder melt. Once melt is seen, *move the tool back and forth on a square pinned device (ICs) or in a circle on devices with round leads (resistors, capacitors, etc) and surface mount pads*. Begin vacuum extraction while moving the tool as described. Once all solder is removed, lift the tip but continue to apply vacuum for a second or two. Not keeping vacuum applied can cause solder to get lodged in the tip and it will require cleaning.

The vacuum action has one other advantage; the air drawn into the tool is not heated. Therefore, it cools the pad quickly and decreases the chance for damage.

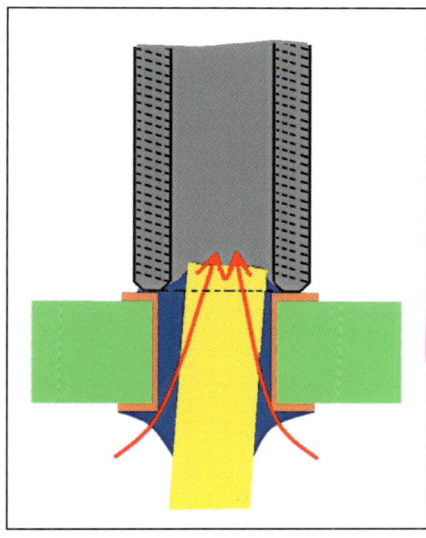

Figure 6-51
The constant air flow draws all of the solder into a collection chamber.

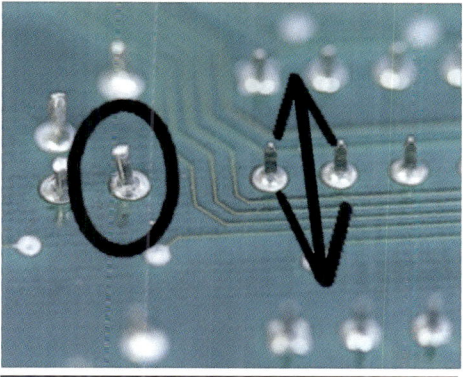

Figure 6-52
Continuous air devices are used in different ways on square and round pins. The motion is back and forth on square pins and circular on devices with round pins.

113

Chapter 7
Removal and Installation
Of Surface Mount Components

The majority of PCBs currently manufactured feature surface mount components. As such, it is necessary to understand their removal and installation. Their size presents unique problems to individuals doing manual rework, repair, and manufacture.

Mastering surface mount installation is easier than mastering through-hole work. If you followed the chapters in this book in the order presented, you will have learned all of the motor skills necessary for surface mount devices (SMDs). You will also have trained your eye to recognize good work. Once these traits are ingrained, learning new procedures follows naturally.

Because of their size, some kind of magnification must be used to inspect the finished work. Ideally, a binocular microscope of X20 magnification should be used for inspection of surface mounted devices installations. As a minimum, inspect your work under X10 magnification.

Surface Mount Component Types

A surface mount device is any electronic part that mounts onto the surface of a PCB and does not use a through-hole connection. Because no drilling is required, lower costs over their through-hole counterparts can be realized. Surface mount components are also considerably smaller. The area of a board's surface is limited and valuable: these space saving devices maximize the useable area.

There are as many variations in the physical layout of SMDs as there are with through-hole components. The defining characteristic of SMDs is their size. Some of the more common configurations are seen in **Figure 7-1**. The dime offers a sense of the scale.

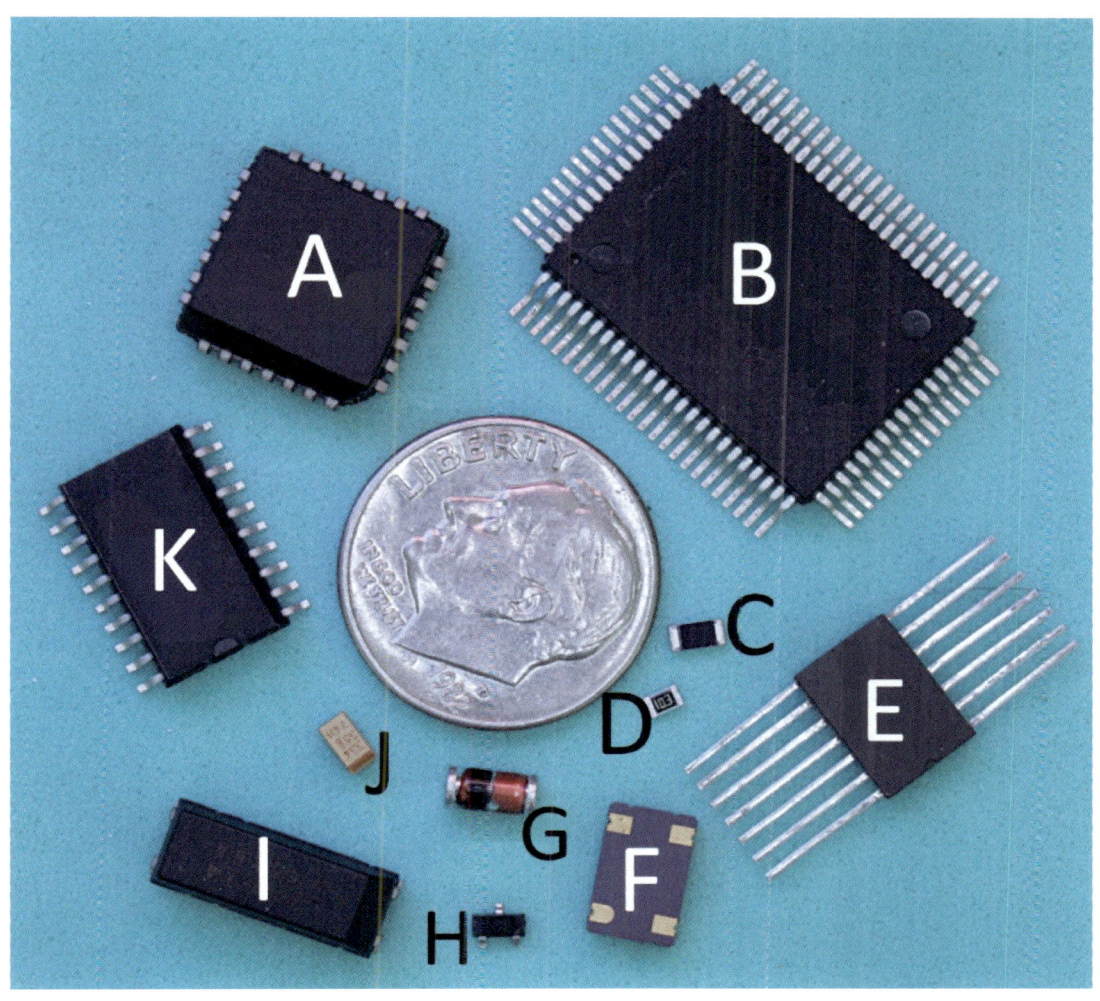

Figure 7-1

A - PLCC (Plastic Leaded Chip Carrier), B – QFP (Quad Flat Pack) can have as many as 376 pins, C – 1208 Chip Resistor, D – 0805 Chip Resistor, E – Flat Lead Small outline IC, F- LCC (Leadless Chip Carrier) viewed from the bottom, G – 1812 MELF (Metal Electrode Leadless Face), H – SOT23 (Small Outline Transistor), I - another LCC seen from above, J - Chip capacitor (a tantalum capacitor as distinguished by the brown line at the 5 o'clock position) and K – SOIC (Small Outline Integrated Circuit, also called a SOP, Small Outline Package).

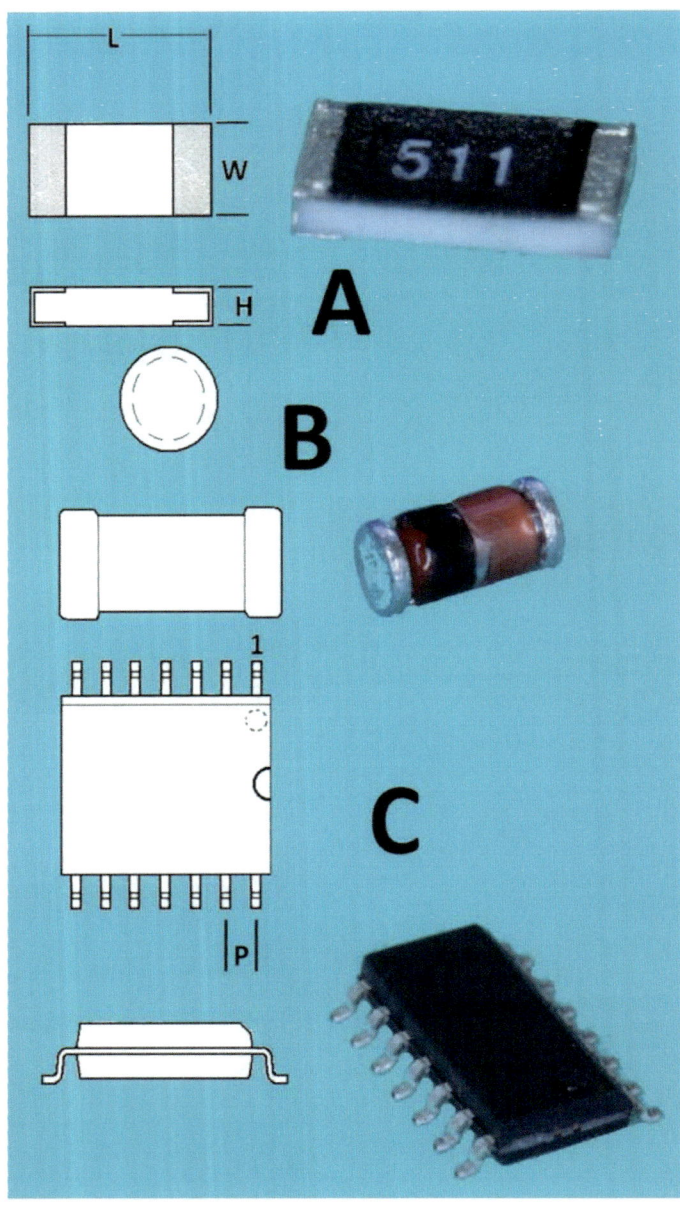

A - Chip resistors are classified by resistance and size. Some may have numbers stamped on them. They *usually* indicate resistance. The first two numbers are the first two digits of the resistance value. The third digit indicates the number of zeroes to add to the first two digits. The 511 indicates a 510Ω resistor. This one has a length (L) of 0.012" and a width (W) of 0.006". You can see the measurements are in mils or 1/1000th of an inch. The resistor is therefore called a 1206. Metric resistors are sensibly measured in millimeters. A chip resistor with a metric size of 3216 is 3.2mm by 1.6mm. This resistor is equivalent in size to the 1206 discussed above. Chip resistors range in size from 01005 to 2512 using Imperial measures or 0402 to 6332 in metric measure.

The black material is the actual resistive compound and should always be soldered into the circuit facing up to allow heat dissipation. Chip capacitors may look much like the chip resistors but will tend to be brownish or tan in color.

B - The MELF is measured in the same fashion as a chip resistor. MELFs can be resistors, diodes, or capacitors.

C - The next component is the SOIC. Pin 1 is marked by a dot or notch on one side of the part. The pins are counted in a counter-clockwise direction (as viewed from the top). Pitch is the distance between pin centers – this one has a 50 mil (1.27mm) pitch. Pitch can be as small as 16 mils (0.406mm). The leads are called gull wings because of their shape. Gull wings are a very common lead configuration for SOICs. The SOIC can have as many as 96 pins.

Figure 7-2

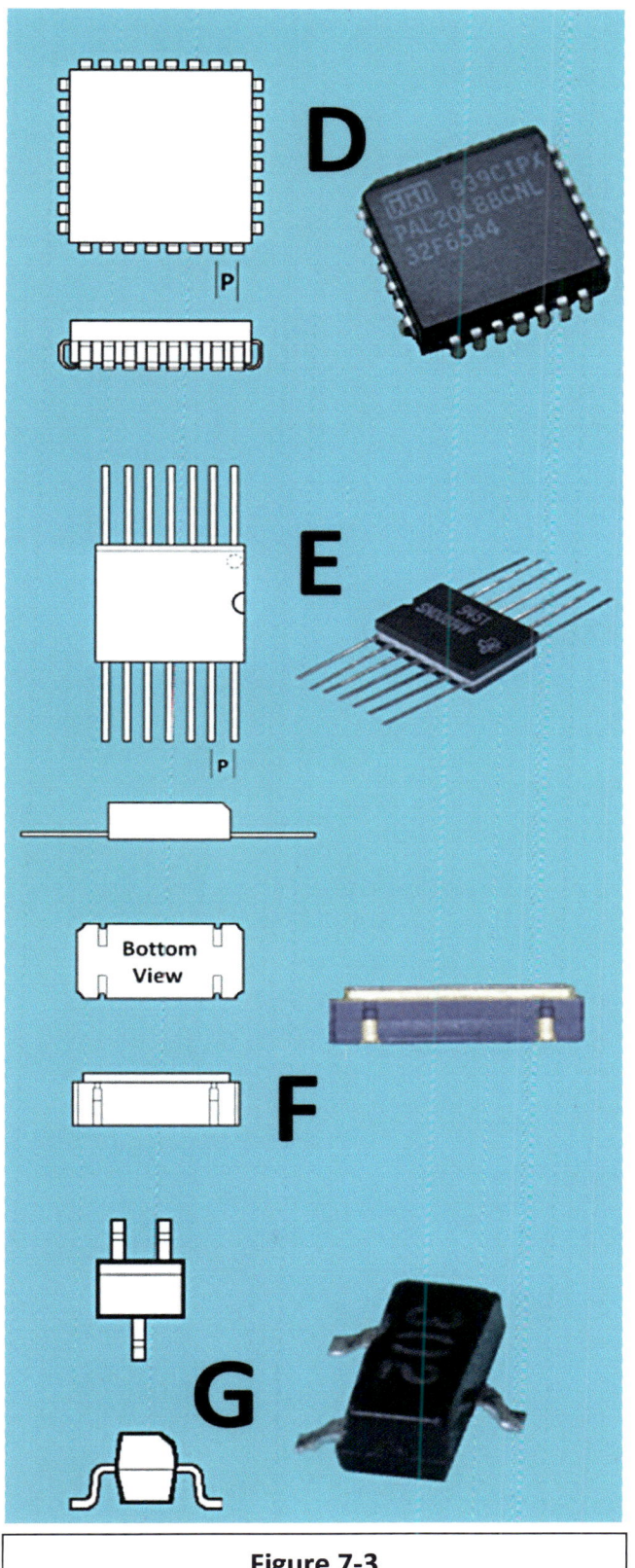

D - Here is a PLCC. The lead configuration is called a J-lead due to the J shape of the contacts (see the right side of the side view drawing). Pitch for this item is 50 mils. The part can be square (varying in size from 0.39" to 1.3", 9.9mm to 33mm) or rectangular with the long axis no bigger than 0.590" (15mm). The component may also be ceramic in which case they are called a CLCCs. This device never has more than 100 pins.

E - Flat lead packages may have leads on two or four sides. Pitch is always 50 mils (1.27mm). The notch or circle indicates the pin #1 position. Pins are counted in a counter-clockwise direction when viewed from the top of the IC.

F - LCC have lead pitches of 20 (0.5mm) to 50 mils (1.27mm). The square LCC can be just over two inches on each side. The leads may have *Castellations* – a semicircular cutout in the leads.

G - An SOT-23. This configuration can be a transistor or diode and has gull wings. There are other package sizes but they all share the same lead layout. Other sizes include the SOT-89, SOT-143, and SOT-223.

Figure 7-3

Installation and Removal of Chip Resistors, MELFs, Chip capacitors, and SOTs

Chip Resistors

Chip resistors, chip capacitors, and MELFs can all be installed using the same process. Clean the pad and the component thoroughly. A soldering iron with a 1/16th inch (1.5mm) chisel tip works well for the larger surface mount resistors. The 0402 (1005 metric) resistors would be better served with a 1/32nd inch (0.75mm) soldering tip. Be sure to use a small diameter solder, 0.015" (0.4mm) diameter is best. Using larger solders will deposit too much material on the pad. All of the solder will then have to be removed and the pad requires cleaning and tinning again. Add flux to one pad and apply some solder to that pad (**Figure 7-4**). The amount of solder necessary will vary by the size of the pad. Getting correct amount of solder will take very little practice. Once the pad has solder, remove any flux residue. Flux should never be used twice because it loses its effectiveness after one heating cycle. Add fresh flux to both pads and align the chip resistor over both pads. One side will be elevated because of the pad you soldered earlier. Patience is needed to align the resistor accurately. Make sure the lettering on the resistor faces upwards. This lettering is always on the dark side of the component. You want to place the lettering facing up so the part can dissipate heat while operating. The black material is the resistive portion of the chip. Once you have the component in position, use an orange wood stick or toothpick to hold it in place. Apply as little pressure with the stick as possible. Excessive pressure will be retained in the part after it is soldered. Be careful not to shift the resistors position. Carefully apply the iron to the point where the resistor and

Figure 7-4
Place flux on a single pad and add solder to create a small mound.

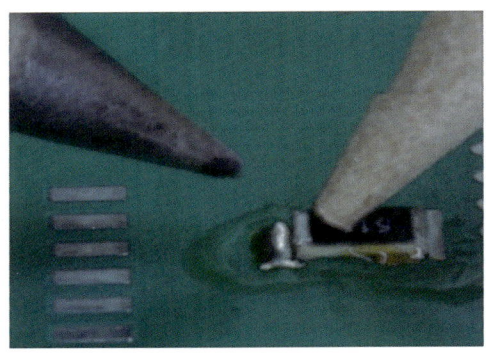

Figure 7-5
Hold the chip resistor down and solder one side – be sure to remove the flux from the previous step.

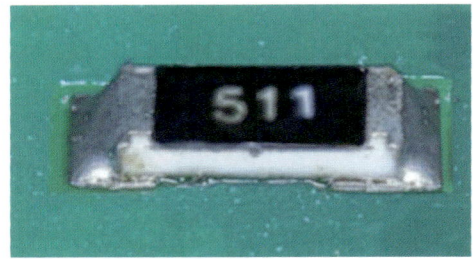

Figure 7-6
An installed surface mount resistor.

solder meet (**Figure 7-5**). Within two seconds or so, the solder will melt and the resistor should drop into place. Once again, inspect the work. The resistor should not have rotated and be more or less centered on and between the pads. Class 3 standards allow a maximum of 25% overhang (based on the components width). For a 1206 (3216 metric) chip resistor, the misalignment cannot exceed 0.0015" (0.038mm). If the alignment is off more than this, the part must be desoldered and the process repeated. If the chip passes muster, solder the opposite side. Place the iron at the junction of the pad and chip, watch for the flux to activate (it will smoke and boil away) and then add a little solder. Wipe the iron off the pad parallel to the board and along the components short axis. Once the resistor and work area are clean, inspect the chip again. The resistor should be centered; the solder clear and smooth, and concave fillets should be visible on both sides (**Figure 7-6**).

Figure 7-7

Jumping from one side of the part to the other is a basic and effective way to remove it.

Removal of chip resistors can be done with a soldering iron having a 1/16" (1.5mm) tip. This method requires the technician to jump from one side of the resistor to the other. Tin the solder tip lightly and apply flux to the pads to improve heat transfer. Good solid contact must be made each time and care taken to avoid too much downward and lateral pressure. After going back and forth a few times, small movements of the component may be seen.

The part may not be completely desoldered even with this movement. Moving too fast and applying pressure can cause a pad to lift and leave the technician with two problems instead of one. Make sure complete solder melt is evident. After bouncing back and forth four or five times, the chip should come off the pads easily. It is possible that the chip will stick to the tip of the iron because of the surface tension of the solder used as a heat bridge.

Another approach involves using two soldering irons. The technician applies both irons simultaneously to opposite pads of the resistor and waits for melt. This requires a degree of virtuosity and may not be for everyone. It is easy to slip and consequently damage other components or the board itself.

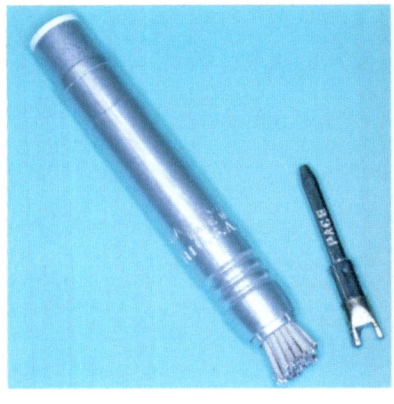

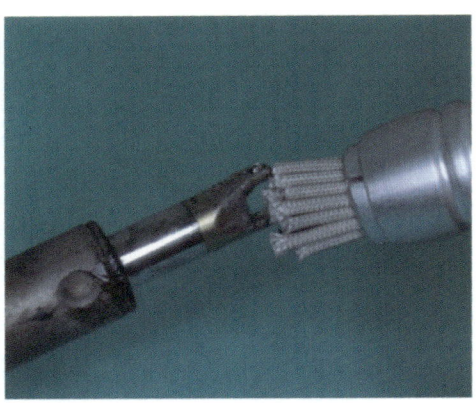

| **Figure 7-8** | **Figure 7-9** |
| A fiber cleaning tool and a bifurcated tip. | Image of tip cleaning. |

Bifurcated soldering tips are also available but seem to be losing popularity. **Figure 7-8** shows a fiber cleaning tool and a bifurcated tip. The tips are sized for the chip they are to remove. The tip shown is for 1206 (3216 metric) resistors and would be ineffective with anything larger or smaller. The fiber tool is used to remove solder and flux from the tip since a sponge would not work. Old solder and flux should always be removed each time this tool, or any soldering tool, is used. **Figure 7-9** shows how the cleaning tool is used. Always brush gently, too much vigor and you may launch hot solder onto a neighbor.

To use the bifurcated tip, apply enough solder inside of the tines to create contact points on the solder pad. However, too little solder and you won't contact both pads, too much and solder will fall onto the board. **Figure 7-10** shows the correct amount of solder in the tip and its application to a chip resistor. Again, be careful not to twist the part prematurely. When the solder finally melts, usually two seconds or so, the part will come free and surface tension will cause it to adhere to the tip for easy removal.

The tip can be used for installation as well but this should be left to someone with a little experience. To

Figure 7-10
A surface mount resistor being removed using a bifurcated tip. Notice that there is sufficient solder inside the tines to contact both pads of the resistor simultaneously.

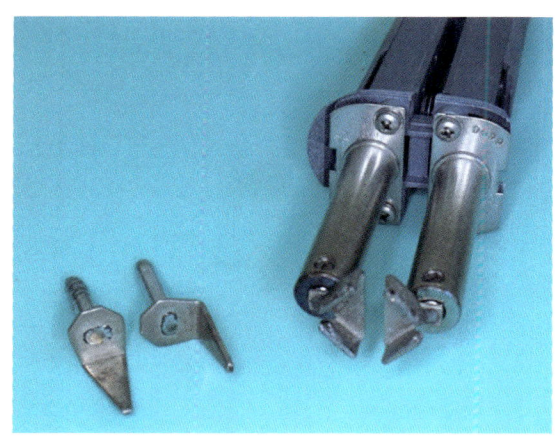

Figure 7-11
Thermal tweezers with interchangeable tips

install a chip using this tip, clean and flux the leads to be soldered. Use an iron to place sufficient solder on each pad to create a good connection. Clean again to remove the used flux and add fresh flux. Align the SMD on the pads using a non-destructive tool. Clean the tip but con't place any solder on it. If you do, you will end up with too much on the pad. Apply the tip as perpendicular to the board as possible, bring the tip down so a tine lands on each pad and straddles the component. Within a few seconds, the solder will melt. Lift the iron straight up. The component should be set almost perfectly on the pads. The bifuracated tip has the advantage of improved centering of the part; since both pads melt simulatneously, the chip self aligns because of surface tension. Remove the old flux and inspect under magnification.

Thermal tweezers are another tool used to remove chip resistors. The tweezers, seen in **Figure 7-11**, have interchangeable tips. The tips to the left are for chip resistors; the tip mounted to the tool is used to remove QFPs et al.

MELFs and Chip Capacitors

MELFs and chip capacitors would be installed in the same fashion as chip resistors. A MELF will be the more challenging because it has the nasty habit of rolling off the board. Many technicians and engineers insist MELF actually stands for Many End up Laying on the Floor. Chip capacitors present a unique requirement; they must be preheated before soldering. Chip capacitors are basically several thin metal plates separated by an insulator. The metal plates have a different rate of expansion than the capacitors insulator. The insulator may be ceramic, tantalum, mica. or a host of others materials. The dissimilar expansion

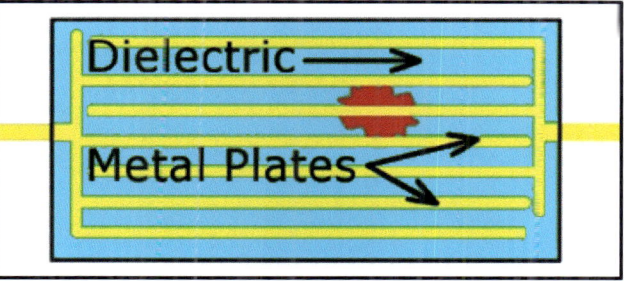

Figure 7-12
Structure of a surface mounted capacitor. A large and sudden heat change can cause the metal to expand faster than the dielectric. This will eventually lead to component failure.

and contraction rate may cause fatigue cracks in the capacitor if it is heated or cooled too fast. The failure may not be evident immediately. As the component is subjected to on and off cycles, the expansion/contraction pattern continues. Eventually the component will fail. To prevent huge temperature variations from occurring during soldering, chip capacitors must be preheated. A heating plate is used that can be programmed to rise to a given temperature at a set rate of temperature change. An increase of 1.8° to 3.6°F (1° or 2°C) per second is common. When the chip capacitor has reached the set temperature, the chip is moved to the PCB board to be soldered into place. To keep the chip from cooling too much between removal from the plate and soldering to the board, be sure to have one of the pads prepared to receive the chip, i.e. have a small mound of solder on it. Be sure to flux the other pad also. The procedure from this point is identical to that of mounting a chip resistor. Removal of MELFs and capacitors is the same as that of a resistor.

Figure 7-13
Solder the point marked A first. Note the fillets on all three leads.

SOTs

Small-outline transistors (SOT), also called TO236ABs, are another group of SMDs and are relatively simple to solder devices. They can be installed like chip resistors. Begin by adding flux to all three leads and pads. Solder point A in **Figure 7-13** first. Because the leads are so small, use solder sparingly on the pad. Again, 0.015" (0.4mm) or 0.02" (0.5mm) diameter solders are preferred. Another option, perhaps not as desirable as adding fresh solder from the roll, is to flux the pad, lead of point A, and add solder to the iron only. The solder from the iron, when placed in contact with the pad and lead, will provide sufficient solder to make a connection. If you chose this approach, do not leave the solder on the iron for more than a few seconds. The solder will deteriorate if heated too long. Double-check the alignment before soldering the leads on the opposite side.

Removing a SOT can be done using the jump technique. You will need a chisel tip 1/16" wide. This size provides sufficient width to contact both pins opposite point A simultaneously. Use solder wick or some other desoldering tool to remove any residual solder. You remove the old solder not just because it is old but also to prevent *bridging*. Bridging is the accidental connection of two points by solder. This accidental connection results in a short circuit and a component malfunction.

Installation of QFPs, PLCCs, and SOICs

All of the chips in this group are installed using very similar methods. What is amazing is the number of ways that these devices can be installed. The use of a soldering iron and a 1/32" (0.75mm) conical tip is one and will be illustrated first on an 80 pin Quad Flat Pack (QFP). Installation of PLCCs and SOICs will follow using paste (second method) and a Pace Miniwave tip (third method). There are several more ways to accomplish installation but these three will suffice as an overview.

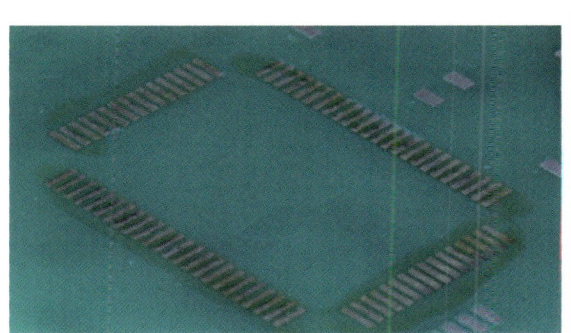

Figure 7-14
Flux the pads before placing the chip on the pads and then add a little more with chip in place.

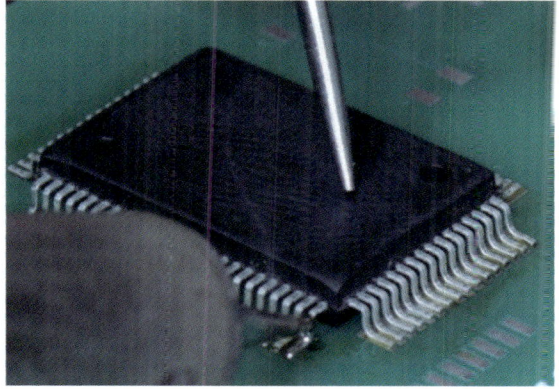

Figure 7-15
Hold the IC down and tack opposite corners.

The hardest devices to install with a soldering iron are QFPs and PLCCs. The correct amount of solder and a very steady, trained hand are necessary. As with every type of soldering, clean the pads and the component leads. Next, place flux on *all* of the pads. Place the chip on the board and align it precisely. It is exceptionally difficult to do this correctly without the use of some type of magnifying device. Each lead must align properly – Class 3 soldering allows an overhang of 25% or 0.02" (0 5mm), whichever is less. As an example, some QFPs have nominal lead widths of 0.22 mm. This would allow a misalignment of 0.055 mm or about 0.002"! With the chip aligned, add a little more flux to the leads and place a very small amount of solder on the iron's tip. While holding the chip down to prevent movement, tack a corner. Contact the junction of the pad and lead. Solder will flow from the tip of the iron to the juncture. When you see the flux activate, drag the iron to the toe of the lead and remove it. The solder joint does not need to be perfect at this point; you are just making sure the alignment is correct. However, do it right the first time, and the connection will not have to be reflowed. If alignment is still good, tack the opposite corner. Check alignment again. The leads are so small that they are easily twisted. These pins can be bent quite easily. If the chip is properly placed, solder the remaining leads.

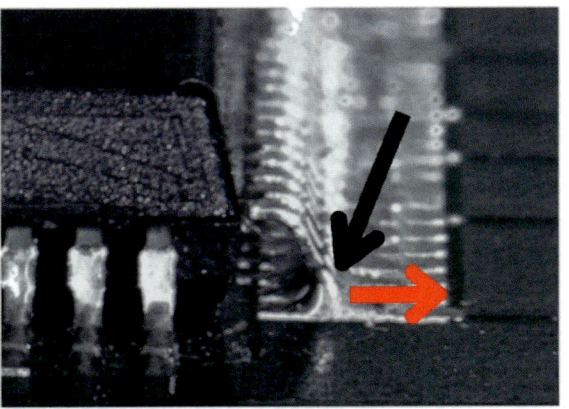

Figure 7-16	**Figure 7-17**
Tacking makes a temporary connection. The joint must be re-soldered to meet specs.	A completer QFP with gull wing leads should look like this. Fillets are visible at the toe and heel. To solder, place the iron on the pint indicated by the black arrow. When the flux activates, drag the iron on the lead in the direction of the red arrow.

You can do this in the same way you tacked the corners, a little solder on the tip of the iron placed at the juncture of the gull-wing lead. Skipping around pins is not necessary

The indicator of a correctly solder connection is a fillet at the heel and toe of every pin. The solder should be smooth and with no bridging. Should there be too much solder, use solder wick to remove the excess, clean the connection, add fresh flux, and solder the lead again. Solder can go halfway up the lead on the toe side and may extend past the upper bend (next to the chip body) but the solder must not touch the chip body.

SOICs and PLCCs

These two device families are installed just as gull winged flat packs. Opposite corners are tacked to facilitate correct alignment. Once the alignment is correctly set, the individual leads can be soldered. The difference is in one standard: the fillet height must not go over ½ of the length of the J-lead on the toe side. The solder may extend to the top bend but it cannot touch the components body on the heel side. Wherever solder has flown, fillets should be seen.

Figure 7-18
Tacking a PLCC before permanent installation. Remember to flux the pads and the leads and make sure the chip doesn't move while being tacked.

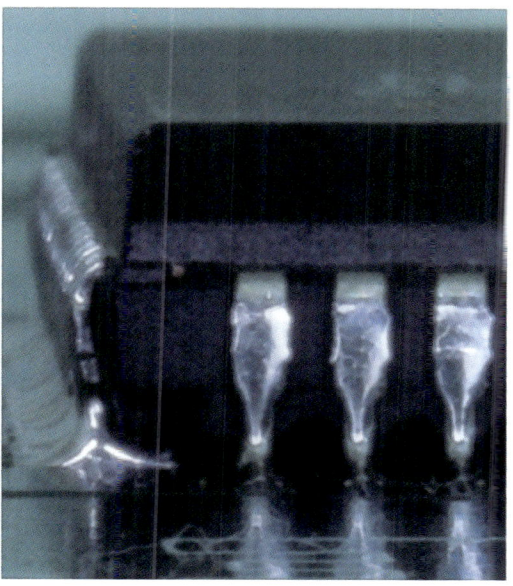

Figure 7-19
Close-up of a properly soldered J-lead on a PLCC.

Specialized installation tools can make chip installation much easier than using a soldering iron only. The use of paste solders and hot air tools is one of these approaches. A hot air tool is seen in

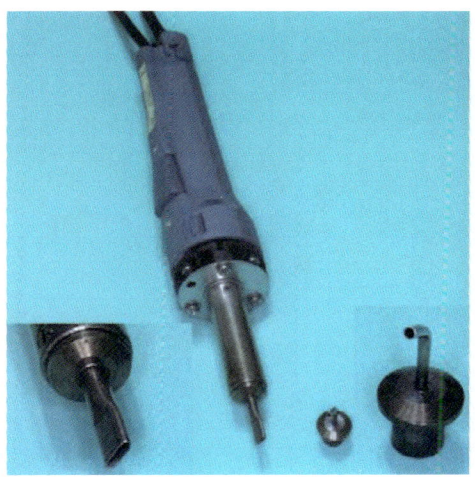

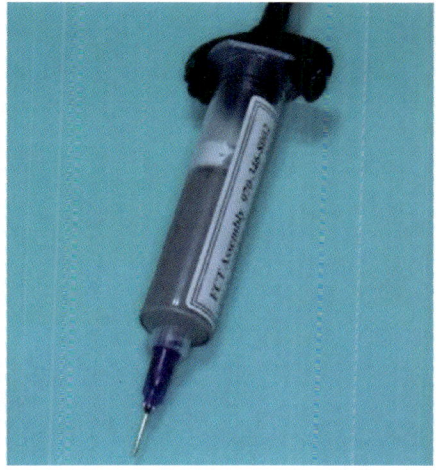

Figure 7-20. Hot air whose temperature can be set from a base unit, is blown out of a hand piece which has interchangeable tips. Two tips are shown but many other sizes are available. Solder paste is used in lieu of rolled solder with hot air devices. The paste comes in several types: many selections of lead free

Figure 7-20
A thermal hand piece and two of many available tips.

Figure 7-21
Solder paste with dispensing attached to a pneumatic device. The tips are also

solder, lead containing solder, silver solder, no clean, water soluble, and RMA. The solder paste is pre-mixed with flux.

One of the unfortunate characteristics of solder past is that it has a shelf life and stringent storage requirements. Solder paste has a maximum one year shelf life and must be refrigerated, but never frozen, between use. It must also be allowed to warm up after it is removed from cold storage.

Dispensing of solder paste can be done manually (plunger in a syringe) or through a powered system (a switch activates a pump whose air pressure pushed the paste through a needle). The paste is sold in tubs or syringes. The syringe has tips of various diameters to help in adjusting the rate at which paste is dispense

Using solder paste is straight forward aside from the special storage requirements. Clean the pads and part. The paste can be dispensed onto the pads first and then the part installed on top of the solder or the part can be installed and then the paste dispensed on the leads. Solder does not have to be placed carefully on the pads alone. A long bead of paste can be dispensed along the length of the leads. The solder will draw to the leads and pads when heated (because of surface tension) clearing the gaps between the pads. The caveat with this is that too much solder can be dispensed. **Figure 7-22** and **Figure 7-23** show the paste being dispensed onto the pads of an SOIC and the paste being soldered. A regular soldering iron can also be used with excellent results.

Figure 7-22	**Figure 7-23**
Paste is being dispensed on each pad.	The first pin in this image has already been soldered. To finish the rest, the technician slowly moves down the line of pins while observing solder melt. Temperature and the air pressure must

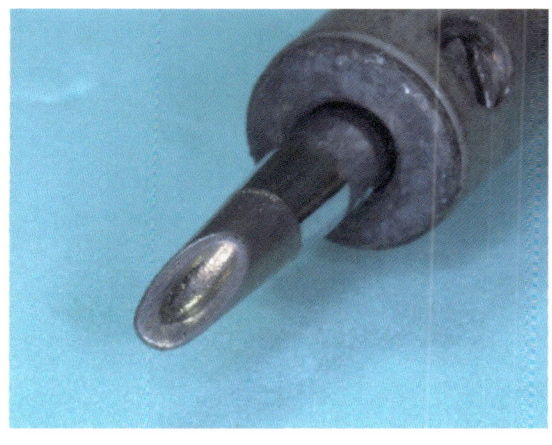

Figure 7-24
An empty horseshoe (Miniwave) tip.
These can be cleaned on a sponge or with
the fiber tool.

Figure 7-25
A Miniwave tip loaded with solder.

Drag soldering is also used by technicians to mount components. Drag soldering involves dragging the iron slowly down the leads of the IC while feeding solder to the point between the iron and the IC. This is a poor method to use. Too much solder is left behind which must be removed with solder wick. Bridging of the leads is certain to occur.

All work should be done correctly the first time in Class 3 soldering. The reheating of pins to remove excess solder will be followed by adding more solder to achieve the proper fillet and solder quantities. That means there were three heating cycles which is two too many. Rework should always be kept to a minimum.

Because this is a gull wing piece, solder can go halfway up the lead on the toe side and may extend past the upper bend (next to the chip body) but the solder must not touch the chip body. Overhang is 25% or 0.02" (0.5mm) whichever is less.

Figure 7-26
Drag soldering using a Miniwave.

Another tool for surface mounted IC is the horseshoe tip (originally the Miniwave by PACE Inc). This tip turns the rule against drag soldering on its head as will be seen shortly. These tips, which come in several sizes suitable for even fine pitch tasks, have a small trough in the tip to hold solder. The solder stays in the trough via surface tension. The Miniwave tip recreates a solder pot with the ability to be turned upside down. **Figure 7-24** shows the horseshoe tip empty. To use the tool, add solder to the tip until a slight bulge is seen, as in **Figure 7-25**. Have the work area clean and flux as you would with any soldering task. Tacking the corners with 1/16" (1.5mm) or smaller chisel tip is a good idea (but not necessary if you can keep the part in position). Beginning at one end of the part (a wide SOIC in this illustration, **Figure 7-26**) drag the iron slowly down the length of the IC. Capillary action and surface tension will draw just the right amount of solder from the horseshoe tip to the pads and leads. Bridging should not occur when this tip is properly used.

Removal of QFPs,

PLCCs, and SOICs

Removing surface mounted ICs of any type is almost impossible (with the exception noted below) using a soldering iron alone. Each pin must be completely cleared of solder and even with the use of solder wick and vacuum tools, some residual solder always remains. If the technician tries to pull the chip off the board in this condition, a lifted pad or broken lead is inevitable. Specialized tips and tools are highly recommended.

The exception to the solder iron alone removal process involves the use of special alloys that are packaged like solder. These alloys are painted on the leads of surface mounted IC that is to be removed. The alloy begins to draw the solder from the connections. It has a low melt temperature (about 300°F, 149°C) and remains in a liquid state for an extended period. By moving the iron around the edges of the

IC, all two or four sides can be molten simultaneously. Once all are molten, just lift the chip, and clean the pads using solder wick and/or vacuum tools. These products are an excellent alternative to the more costly dedicated surface mounted IC removal tools. Low volume work is where this method shines. Two of these products are Chip Quik® and LowMelt®.

IC removal at higher volumes would be better served with specialized tools, one of which is seen in **Figure 7-11** and shown being used in **Figure 7-27**. These tips come in every common IC package size on the market. Using these tips is just like using a soldering iron. The tip temperature will have to be adjusted slightly higher or, if a soldering station is used, an *Offset* will have to be input. Offsets are settings that compensate for tip mass. The larger the tip, the greater the heat radiated into space. If offsets are not input, the tip temperature will be lower than indicated. Once the temperature is set, add solder to the tip where it will contact the IC leads. Flux should also be added to the IC to facilitate heat transfer. Place the tip vertically on the IC, wait 2 or 3 seconds, and then *gently* twist and lift. If you try to twist and the IC doesn't move, stop. Heat the leads for a few more seconds and try again. The IC should come off easily. These tips do leave solder and flux residue on the board, which will have to be removed.

Figure 7-27	**Figure 7-28**
A SOIC being removed using a dedicated removal tip.	The same IC pad as in the previous figure with the IC removed. The pad has not been cleaned to show residue left behind.

Mixed Technology Boards

A great number of PCBs have mixed technology – they contain both through-hole and surface mount devices. To solder these boards at manufacturing plants, the surface mount devices must be glued onto the boards. The glue can be seen as the red material in **Figure 7-29**. Glue necessitates a slightly different approach to SMD removal. The technician will have to spend a few extra seconds on the chip leads to overcure the adhesive. Once overcured, the devices can be removed normally. Using the adhesive for reinstallation of the part is determined by the manufacturer or your company's requirements. The use of these adhesives, if not prohibited by you company, can make soldering during the installation much easier. You can align the part accurately, wait for the glue to cure, and then solder the piece into place.

Figure 7-29
The surface mount components on a mixed technology board are held in place by glue (the red dots) during the wave soldering process.

Ball Grid Array (BGA)

The Ball Grid Array (BGA) is an interconnection technology that is becoming one of the dominant styles in the electronics market. **Figure 7-30** Shows a picture of a BGA. It is easy to understand how the part got its name. The connections are arranged in an array of balls. This IC, with the solder balls attached, can then be placed on a circuit board, which has an identical array of pads or semi-circular troughs, to receive the balls. The balls themselves are usually a low melt temperature solder and are quite small, 0.30mm to 1.27mm (0.012" to 0.05"). These balls rest on a high temperature solder alloy with a semi-circular cutout. In most applications, the balls need to be soldered to the IC only. There are instances in which the IC and PCB are soldered together. Both connections can be repaired but only the former will be discussed.

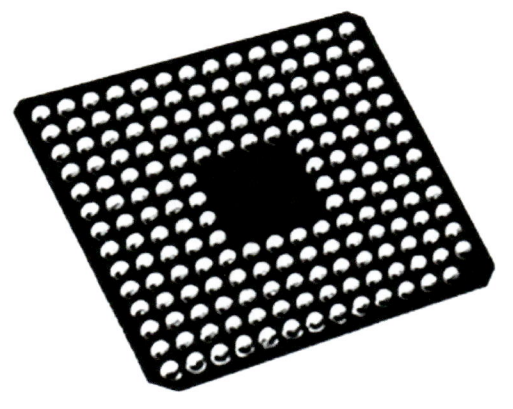

Figure 7-30
A BGA. The balls can be seen "sitting" on the ICs contact pads.

Preparation

Repair of a BGA begins with de-balling the IC. Apply flux to the solder side of the IC. A high-tack flux, one that is not thin and runny, is easiest to use. Solder paste is an example of a high tack solder. Removal of the balls can be done with a wide soldering tip and some solder wick. Place the tip on and the wick on the solder side of the IC and slide it across the IC. The solder should readily flow into the wick. Work at a pace that allows good solder wicking but prevents excessive buildup of heat in one spot. Clean the de-balled IC with an approved solvent (isopropyl alcohol is often used) and then clean it again with de-ionized or distilled water. De-ionized water is cheaper than distilled water but any organic residue that may have been in the de-ionized water will still be present. Use the water just like a cleaning solvent. Brush it on and let dry. Do not soak the chip in the water, it may delaminate when re-installed (the absorbed water will expand during the installation process). This may cause the IC to pop like a popcorn kernel. This is why the effect is called **popcorning**. The chip must be inspected for *coplanarity* before it is installed. Coplanarity means that the chip is straight and level on the solder side of the IC.

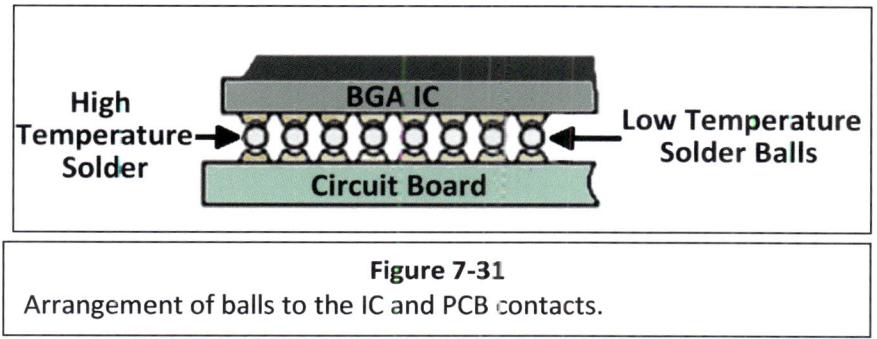

Figure 7-31
Arrangement of balls to the IC and PCB contacts.

The skill (and cost) in BGA repair come into play during the re-balling process. Many methods of re-balling are available. Some approaches, however, may not meet the highest conformity standards. Two high reliability methods will be discussed. High reliability work will normally be inspected using X-ray machines.

Re-Balling Using Solder Paste

Re-balling with solder paste is a simple, inexpensive, and quick way to repair a BGA. Minimal equipment is needed and a well-trained technician can produce excellent results.

Once the BGA is prepped as described above, preheat it with an oven or hotplate. Lay the BGA in the

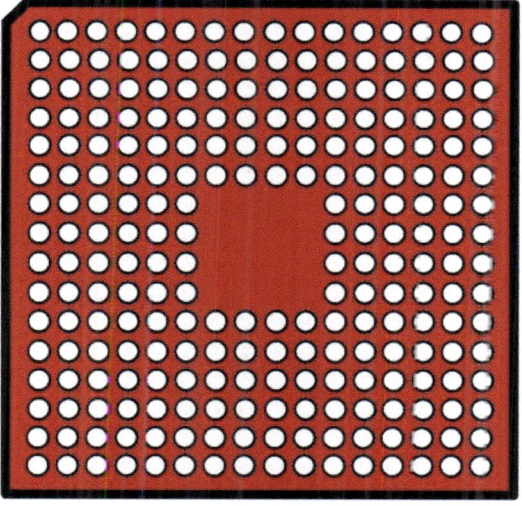

Figure 7-32
A stencil (*also called a grid, template or preform*) would be placed on the contact side of the BGA to align the solder paste. The grids are also available for re-balling using spheres. The cut corner in the upper left marks pin 1.

oven, while the oven is at room temperature, and then turn the oven on. This prevents cracking in the IC that may be caused by thermal stress. Then find a stencil (also called a template) that has the same pattern as the IC. The stencil will be heat tolerant and have openings cut into it, which will align with the BGA pads. Secure the stencil with anti-static tape to the IC. Be sure to get as snug a fit as possible. If the stencil is left loose, solder paste will run into areas where it is not wanted. This may short some of the pins on the IC. Now apply a layer of solder paste to the template. Remove any excess paste. The paste should be flush with the top of the holes in the stencil.

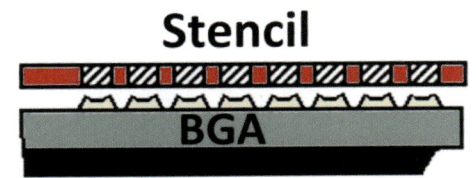

Figure 7-33
The stencil will be placed securely on the BGA. Solder will then be placed in the openings of the stencil. This solder will be melted to form the new balls.

The solder can now be reflowed. Use a hot air tool like the one shown in **Figure 7-20**. The base station, which controls the tool, should be set to about 500°F (260°C). Carefully apply heat to the template and IC. The distance of the tool to the surface of the BGA should be about ¼" to ½" (6mm to 12.5mm). Within 10 or 15 seconds, the paste should begin to melt. Depending on the size of the IC, the soldering should take a minute or less. Be aware that some stencils are not designed to be subjected to extended periods of heat. Heat them for too long and the stencil becomes a onetime use item. Worse yet, some may curl or lift when subject to heat. When the solder has cooled, remove the template and clean the IC with isopropyl alcohol or the solvent recommended by the process designer. Cleaning with de-ionized water is recommended as this removes any residual flux. Inspect the BGA under X20 magnification. You want to make sure all of the flux was been removed, that there are no missing balls, no shorted balls, and no foreign debris (cloth fibers, hair, etc).

Re-Balling Using Solder Spheres

A better method of BGA re-balling requires the use of *solder spheres*. A solder sphere is designed to fit precisely on the contact pads of the BGA. They come in a range of sizes to fit the majority of BGAs on the market.

Start by selecting the correct grid/preform. A fixture or pan may be supplied with the grid so be sure to use both. The pan holds the preform and IC securely in place and offers a way to recover unused spheres. Thoroughly flux the IC and place it face up in the pan. Align the grid over the IC contacts. Make sure the grid is flat in the fixture. If it is bowed, the fixture most likely needs to be cleaned. Once the grid is properly in place, pour the spheres into the pan. Make sure that every hole in the grid has a sphere. The excess can be poured off. The spheres can now be re-flowed using a convective oven or a suitable hot air fixture. A convective oven can ramp the temperature up at a controlled rate minimizing thermal stresses. If a hot air fixture is chosen, one designed for the size BGA being worked on must be used. Once reflow has occurred, let the IC cool.

Remove the grid and then the BGA from the fixture. The IC must be cleaned with a proper solvent and then cleaned with de-ionized or distilled water. Use of a bake-out oven (an oven specifically designed to remove moisture from the IC) may be required. The controlled increase in temperature removes the moisture slowly and prevents the popcorn effect.

This is a generalized procedure. Re-balling supplies usually come in a kit. Follow the instructions supplied with the kit.

Figure 7-34
Place the already fluxed BGA into the fixture, place the grid over the IC and pour the solder balls into the fixture.

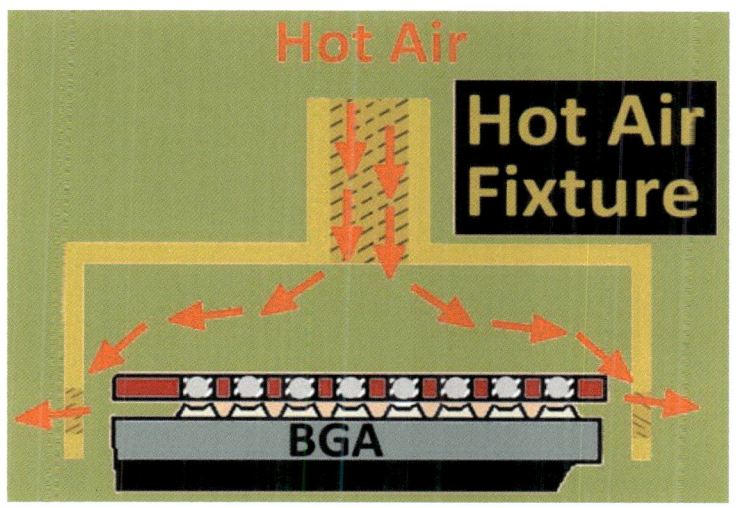

Figure 7-35
A hot air fixture should be matched precisely in size to the BGA.

Chapter 8

Track, Pad, and Board Repairs

Repairing circuit cards is one of the more challenging aspects of electronics. It is also one of the most satisfying. This chapter will deal with the techniques of proper board repair. Some specialized tools are occasionally needed but a majority of the work can be performed without them.

Damaged or Missing Tracks

This section will help you to identify board damage that requires repair and the proper methods to expedite the repair. Track repairs may be tedious but can be accomplished quickly. It is recommended that some type of binocular microscope be used while doing this work. The microscope makes the work much easier. As a minimum, the finished work must be inspected microscopically with a minimum of X10 magnification. Many of the errors made during the repair process cannot be detected with the naked eye.

Lifted Pads and Tracks

Lifted pads may occur due to component overheating, during manufacturing, or during the repair. If measling of the board has not occurred, the repair is quite simple because none of the substrate has to be removed.

Inspect the work to determine the extent of lifting. Make sure the pad is intact. Carefully clean it and the area from which it was lifted. Apply an appropriate adhesive (***NEVER USE CYANOACRYLATE GLUES!***) and lay the pad down exactly where it was originally. Use a tool with a fine point to apply the glue; you want to place the adhesive only where it is needed. To keep the pad in place and make sure it adheres properly, a small Teflon

Figure 8-1
Identify any lifted pads and inspect for board damage.

Figure 8-2
Apply adhesive only where needed.

block should be placed over the hole and held in place by a polymide tape (or clamp). Once the glue dries, remove the block. Excess glue can be removed with the correct solvent or by using a brush attached to a rotary machine. The brush is safer in

some instances because it has no solvents on it that can inadvertently seep under the repaired pad and cause it to delaminate again. A wooden toothpick with some solvent applied to it can also be used to safely remove the excess glue.

If the pad was a through-hole, make sure that the hole is clear. A drill bit can be placed into the holes to clear obstructions. Do not use a rotary tool to open the hole, spin the bit with your fingers only. This will keep the through-hole from being enlarged. Should the lifted pad be for a surface mount component, the repair proceeds in the same fashion.

Figure 8-3
A Teflon block secured with Kapton® tape will guaranteed a properly secured pad.

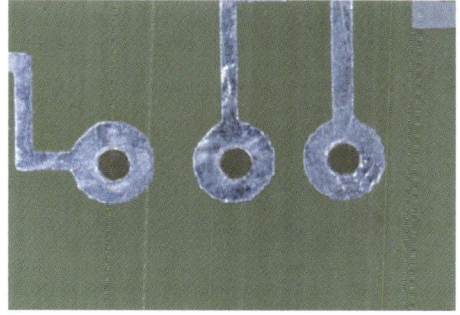

Figure 8-4
Inspect the completed repair and remove excess glues around or in the pad.

Broken or Missing Tracks

Tracks can be replaced using four methods: 1) a track frame without adhesives backing, 2) a track frame with adhesives, 3) using a "busbar" or, 4) by using a "staple."

Using a Track Frame without Adhesive

Track repairs, plated through-hole and non-plated through-hole repairs can be done with the process described here. Track repairs on surface mounted boards can also be done. However, this

Figure 8-5
Two types of frames. One for general purpose use, the second for replacing traces only.

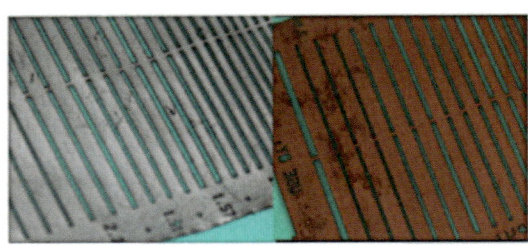

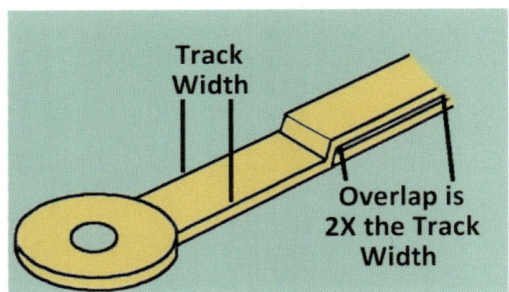

Figure 8-6
The frame on the right is a track pad frame with adhesive on one side.

Figure 8-7
Drawing showing the minimum overlap needed.

Figure 8-8
This eyelet has lifted and shows measling.

method does not work with pad repairs for surface mounted devices. The adhesive track repair approach described in the following section must be used for surface mount pads.

Track Frames, a collection of traces in different widths and shapes come in many styles. They can be tinned, un-tinned, thermally activated adhesive bonding, etc. **Figure 8-5** shows just two of the commercially available track pad frames. The brighter of the two (upper left) has not been pre-tinned and this makes it more difficult to work with when compared to the factory tinned frame. The tinned frame may cost twice as much as its non-tinned counterpart but try to use a pre-tinned frame whenever possible. The ease of use far out ways the extra cost. Before using any traces from either frame, the oxidation must be removed from the pads with a white eraser. Proper cleaning with a solvent must follow.

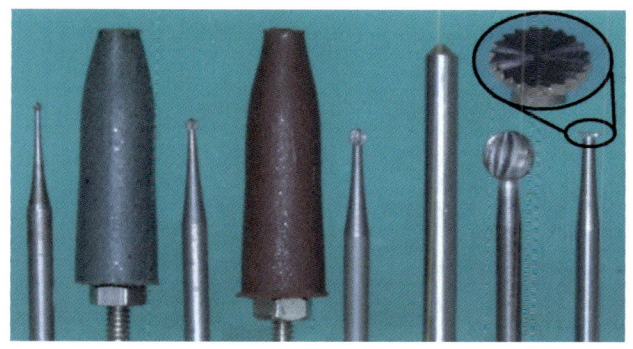 

Figure 8-9	**Figure 8-10**
Shown here are four sizes of ball mills, two bullets used to sand away excess epoxy, a swaging tool (has the conical tip) and a saw.	Three funnelet sizes and one size eyelet. Funnelets are usually used on multi-layer boards. Eyelets are used on single sided boards.

The damaged trace needs to carefully removed; you do not want to damage the base material or, if it is already damaged, cause more damage. Use a hobby knife and score the track to which the pad is attached. Score the track slightly beyond the damage. Don't try to cut through the pad in one stroke. After scoring, grasp the damaged end and move it up and down a few times. The track will break along the score line. You can see in **Figure 8-8** the removal of the damaged pad and a section of the track. On y enough track should be removed to let you accomplish the repair. You will also note that the board is measled. This will require excavation of all of the damaged material. If the damaged area is small, under 1/2"(12.5mm), epoxy can be used as a fill material. Colorants can be added to the epoxy to create a match to the original substrate. Repairs larger than1/2"(12.5mm) will be discussed later.

Figure 8-9 and **8-10** show the tools needed for this repair. A selection of *ball mills* to excavate the damaged substrate are required as well as a *swaging tool* and a small *circular saw* (to create an undercut). These terms will be defined shortly.

The damage has been identified and it is now necessary to remove it. If the measling does not go through all of the PCBs layers, do not go through the board during the repair. Remove only the damage (measles) to the depth necessary. The *ball mills* (bits with serrated circular heads for material excavation) are used for the removal. Insert the largest mill you feel comfortable using into the rotary tool. Don't one that is so large that it removes good material. Set the tool to about 4000 RPM. With the board secured, apply the mill to the damaged area and work in small circle only on the damaged area. Hold the rotary tool firmly while working on the PCB. The mills can grab the board and cause the tip to run across the board. Most rotary tools run in a clockwise direction. For a right-handed person, the tool will want to run away from you: it will run towards a left-handed person. Change to a smaller ball mill for the extremely fine work. You should inspect your progress periodically under magnification to see if all the damage is gone. Adding some cleaning solvent (such as isopropyl alcohol) will make the delaminated areas stand out as white spots against the substrate. When the damage has been removed, use the circular saw (bit with a flat circular tip) to make the edges of the milled area vertical. Do this without using the rotary tool, rotate the saw with your fingers only to keep the hole from enlarging. Next, attach the circular saw to the rotary tool and place the saw blade into the excavated hole. The blade should go below the top of the board when it is placed in the bottom of the excavated area (**Figure 8-11**). Try to be as perpendicular to the board as possible. Using the same tool speed as before, move the rotary tool around the perimeter of the cleared area. If this is done correctly, backlight the board and a halo should appear around the excavated area (**Figure 8-12**). Epoxy will be placed in the hole and get into the undercut. The undercut acts as mechanical bonding for the repair.

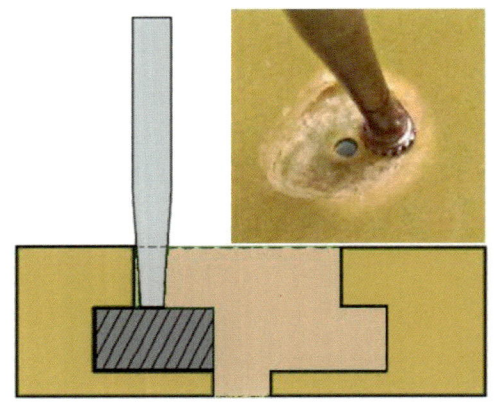
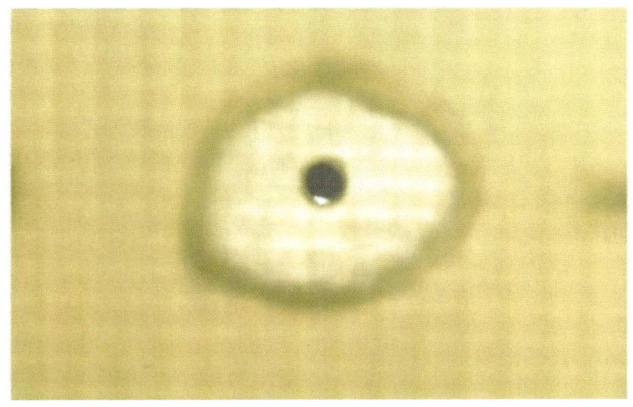

Figure 8-11	**Figure 8-12**
The saw must be perpendicular to the board. The saws shaft acts as a stop to limit the undercuts size.	The halo is easy to see when a light shine through the board. The hole in the board does not extend through the PCB.

The void left in the board must now be filled. As always, use the proper material. Many commercially available epoxies are quite suitable. Epoxy is a two part material, a resin and a hardener, and must be mixed thoroughly. During mixing, try to avoid adding air bubbles to the glue. Air bubbles create voids that weaken the repair. To remove these bubbles, place the glue on a piece of safety glass or plastic. Take another piece of glass or plastic and place it on top of the glue. *Squeeze gently* and pull the top glass off parallel to the lower plate. This forces air bubbles out of the mixture and then you can select the "cleanest" part of the epoxy mixture. Place the glue into the hole using a toothpick; make sure the epoxy gets into the undercut. You can see how the undercut strengthens the repair by acting as a wedge. Make sure the glue forms a mound on the board but covers only the repair area, not adjacent traces or pads. The mound is necessary because most epoxies shrink as they cure. Next, inspect the glued area through a bright light. Air bubbles will appear as black circles and should be removed if they are in the repair area. Bubbles above the repair can be ignored. Air bubbles should make up less than 10% of the repairs volume – zero air bubbles can be achieved with the cautious use of a toothpick. If the repair is of a through-hole pad, place a Teflon block over the hole on the component side and secure it in place (polymide tape or clamp). The block prevents epoxy from running through the hole. If you use something other than a Teflon block, it may end up glued to the board. At this point, follow the adhesive manufacturer's recommendations for cure time. Cure times can be shortened by using a low temperature oven.

An epoxy will have a work time (also called a pot life). This is the amount of time you have to apply it and make adjustments before it begins to set (cure). Five minute epoxies are a good choice if you have done a board repair before. Use a longer work time epoxy if this is the first repair you have done.

Once the glue cures, the excess needs to be removed. The easiest way to do this is by using the abrasive bullets (**Figure 8-9**) in a rotary tool. Two bullets are available. Start by using the green bullet. It is the most abrasive of the two. Be careful not to abrade anything other than the epoxy. Once a majority of the adhesive has been removed, change to the brown bullet. The objective of removal is to get the repaired area *even with the top of the board*. The best gauge you have to determine when you reach this level is your finger tips – variations of a hairs breadth can be felt with them. Be careful not to go below the board. If you do, the area should be excavated and the entire gluing and leveling process repeated.

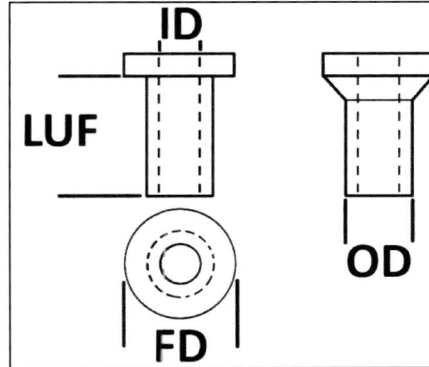

Figure 8-13
An eyelet (left) and funnelet (right) showing measurement points.

The repaired area is now level and you must select an eyelet (**Figure 8-12**) and open the through hole so it can be used again. If the original through-hole was 0.032" (0.8mm) wide and the board is 0.0625" (1.6mm) thick, you need to select an eyelet with a 0.032" (0.8mm) Inside Diameter (ID) and a length under the flange (LUF) greater than the board's thickness to allow swaging. The LUF should be about 0.025" (0.6mm) greater than the board thickness. Also the funnel diameter (FD) must be greater than the

139

Figure 8-14
A repair of a burned area should look similar to this image.

Figure 8-15
Swaging tool with backing plate.

hole to be drilled but not larger than the original pad. Once you have found a suitable funnel, read the outside diameter (OD) from the data sheet. **Figure 8-13** shows where these points are taken from. This will tell you the size of the bit you must use to reopen the through-hole. Once a hole has been drilled in the board, test fit the eyelet. It should go into the hole with slight friction and protrude through the opposite side of the board when the eyelet base is resting on the board.

Select a pad and track from the frame that matches the original pad diameter and track width as closely as possible. Remove oxidation from the track with an white eraser before removing it from the track frame. Test fit your selection with the chosen eyelet. The flat part of the eyelet must be *placed on the component side of the board*. Remove the excess track. As with the lifted track, make sure the replacement track overlaps the original one by at least two times the width of the pad but not more than 1/4" (6.25mm) (go back to **Figure 8-7** for the example). You should have something that looks like **Figure 8-14**. The eyelet extends through the replacement track. Notice that you can just identify the area that was excavated and repaired with the epoxy. Also note that the area is flat.

The next step is to swage the eyelet. Swaging is the process of bending the eyelets walls over to form a cone. Swaging forms the equivalent of an electronics board rivet. Place a small piece of flat metal on the component side of the board to give the eyelet good support. The conical tip of the swaging tools fits into the opening on the trace side of the PCB (**Figure 8-15** and **Figure 8-16**). By applying slightly increasing pressure to the swaging tool and moving it around the eyelet in a circle, the edges of the eyelet will begin to bend outwards. As you keep pressure on the point of the swaging tool, increase the size of the circles you are making. Now take another small piece of metal (usually a small metal disk is on the opposite of the swaging tool) and press down firmly on the swaged eyelet. This will cause the swaged area to flatten. If done correctly, you should see no cracks in the swaged area. Minimum specifications allow two cracks, neither of which radiates into the barrel of the eyelet. In addition, these two cracks cannot be within 90° of each other. Finally, solder the new track down to the old. Soldering the eyelet is optional, as it will be soldered when the component is installed.

Figure 8-16
Swaging tool inserted into eyelet.

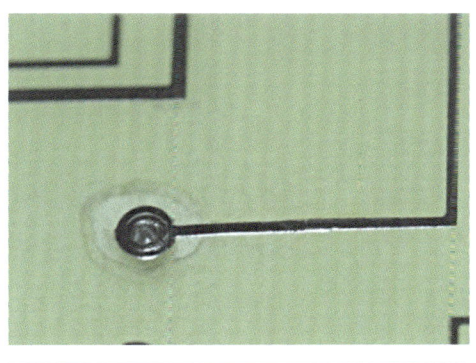

Figure 8-17
A completed burn repair. The image is backlit to highlight the repaired area.

Soldering the track end of the repair can be a little tricky. If the track frames leads are pre-tinned, this step is simplified. Apply flux to the original trace and align the new track with the old. Prepare the correct soldering tip and place a very small amount of solder on it. While holding the track down with a toothpick or orange wood stick, begin soldering where the new track just overlaps the old and draw the iron toward the original (old) track. As you move the iron, the toothpick /orange wood stick should follow. This following motion prevents the new track from coming up while the solder cools (**Figure 8-17** shows a completed repair). A tiny fillet should be seen at the heel of the installed track.

Using a Track Frame with Adhesive

Adhesive backed track frames can be used on al types of repairs. They are ideal for NPT work because an eyelet is not needed to hold the pad down. No special equipment is required.

The type of damage will determine the process needed for repair. If a pad is lifted and measles are seen, the damaged substrate will have to be removed and epoxy filler used. Once your are working with a good base, whether you started with one or had to perform a repair to get a good one, the following restoration method is the same for both.

Begin by selecting the correct replacement pad on the adhesive track pad frame. Remove the oxidation from the tinned side of the pad you selected. Cut the pad from the frame. Cut the pad length

Figure 8-18
The dark green area must be replaced.

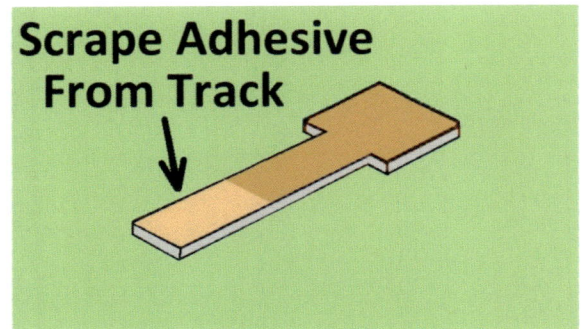

Figure 8-19
Remove adhesive only from the area of the replacement track that will overlap the old track.

long. Fit the pad to the repair area but put it on upside down. Carefully scrape the adhesive off the pad. The adhesive should be removed to the spot where the new track overlaps the old. Cut the new track so that it overlaps the original pad 2 times the pad width to maximum of 1/4" (6.25mm).

Place the track and pad on a piece of polymide tape, tinned side of the pad attached to the tacky side of the tape. Notice that in **Figure 8-20**, the tape covers only the section of the track where the adhesive is still present. Position the tape/pad accurately in the pads final position on the board. The packaging or kit that the adhesive track frame arrived in will have instructions for temperature settings of the heating tool. The tool can be a soldering iron, a lap flow device discussed earlier, or some similar device. To meet Class 3 standards, the temperature must be set in accordance with the instructions. Class 1 and 2 repairs can be done with a 25 watt (or lower wattage) soldering iron.

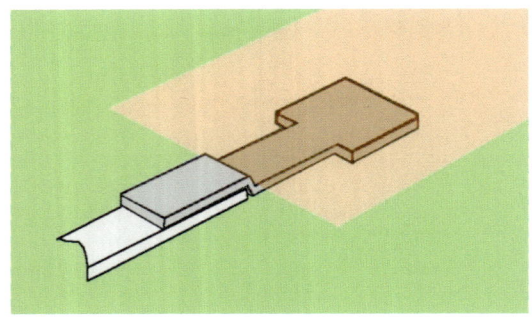

Figure 8-20
Tape down the section to be heated with a polymide tape. The tape also keeps solder from running onto the new track.

Place the iron on the tape/pad (where you didn't scrape the adhesive off) and with light pressure move the iron along the length of the track. After a few seconds, the adhesive will have set. Finally, solder the connection where the track overlap occurs. The soldering methods and standards are the same as the track repair done earlier. There should be a small solder fillet at the termination point of the new track. Clean the pad thoroughly and inspect it.

Figure 8-21
Correct overlap. Flux has been added and the busbar is ready to solder.

Figure 8-22
Busbar being soldered while held down with an orange wood stick.

Busbars

A busbar is a field expedient way to repair a broken trace. It consists of nothing more than a short piece of solid wire. The busbar wire should have the same width as the section of track it is replacing.

First, select a solid wire whose gauge matches the track width. Cut the wire so that an overlap of twice the track width is achieved on both sides of the original trace. To make working with a circular wire easier, you can use duckbill pliers to flatten the wire. Test fit the busbar (**Figure 8-21**).Make sure the trace and busbar are clean, then add flux to both ends of the track. Select a soldering iron tip and place a small amount of solder on it. Holding the track down (polymide tape, toothpick, etc.) and solder both ends of the busbar, Start soldering at the point where the repair overlaps the track and slide the iron toward the original undamaged trace and onto it. This assures concave fillets w ll develop on both sides of the busbar and that the entire overlap is soldered. **Figure 8-23** shows the comp eted busbar.

Busbars are an excellent way to replace damaged traces that had curves in them. Once again, select the correct solid wire to use. Form the wire using your duckbill and round nose pliers. You do not need to flatten the wire. Once you have achieved a god fit, flux ends of the bar and trace then solder as you would a straight busbar. Your result should look like **Figure 8-24.**

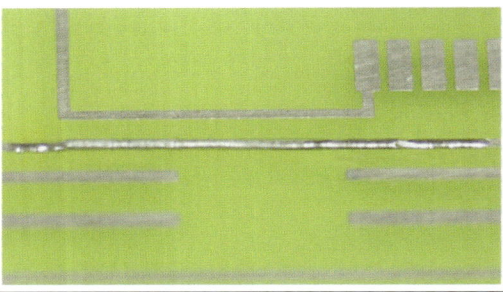

Figure 8-23
A completed busbar. Small fillets are present at either end of the work.

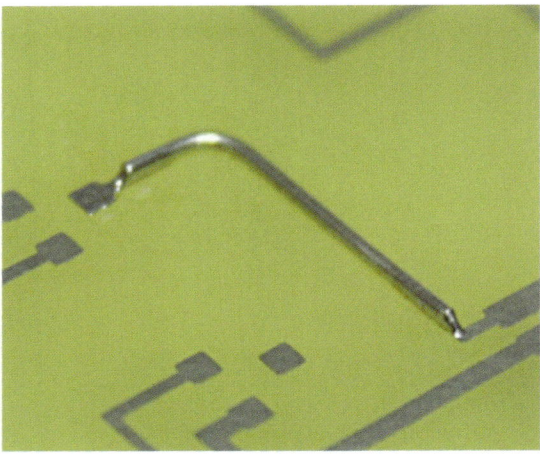

Figure 8-24
A busbar with a single bend – multiple bends are quite easy with this technique.

Figure 8-26
Staple prior to lead forming.

Track frames generally do not have traces with 90° angles. However, you can use a straight trace and fold it over to achieve the turn. The problem with this is the 90° bend. It leaves a weak spot in the trace. To strengthen it requires the use of an adhesive. A small drop of epoxy directly on the bend will minimize mechanical strain. Finally, if you have to use a track with an adhesive already on one side, you will have to form two bends at the 90° point to get the adhesive to face the board as a single bend will leave adhesive facing upwards on one leg of the bend. Multiple bends exacerbate the problem.

Staples

The staple presents a yet another method of replacing a broken trace. The drawing in **Figure 8-25** shows what a staple looks like and how it got its name. It does look very much like a staple. The staple is the most invasive of the repairs we have to work with. Two holes must be drilled into the PCB immediately in front of the undamaged traces. Caution must be taken not to frill into the good portions of the trace or of damaging blind or hidden vias in multi-layer boards. Also, the alignment in front of the traces must be accurate. Misalignment will require you to bend the wire when such a bend was unnecessary (and it is unsightly). **Figure 8-26** shows a set of properly drilled holes with the staple already fitted. The wires must be long enough to allow the usual two track widths overlap. The insulation of the staple on the component side of the board must be flush with the board.

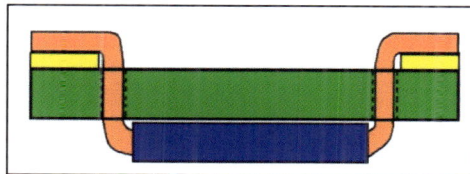

Figure 8-25
Drawing of a staple. The blue area represents the insulation of the wire – this insulation does not have to be used unless specified.

Once you have the wires through the hole, clean everything. Bend the staple wires parallel to the remaining portion of the old trace (use a wooden or plastic tool to prevent lead and board damage). The wire ends of the staple should be in maximum contact with the track. Finally, flux the wire and track thoroughly. Select the correct size of soldering iron tip (width of the pad) and solder. Place some solder on the tip to form a heat bridge, apply the iron close to one of the holes and slowly move it towards the old trace. You may have to feed some solder into the joint to get a proper connection. As with all good soldering, a fillet should appear on both ends of the staple when you have finished. Words of caution, if you leave the insulation on the staple, and you should, make sure not to apply pressure to the insulation while soldering. The insulation will be split by the heated wire and pressure. The final product should like something like **Figure 8-27**.

Figure 8-27
A properly soldered staple – note the fillets at both ends.

Jumper Wires

Sometimes it is necessary to modify a circuit boards connections, not on scale that requires a completely new board, but with the addition of a few new connections. Many of these connections can be done with a *jumper wire*. A jumper wire is a wire soldered onto the connections of a component already mounted to a PCB and whose job it is to modify or repair a circuit card.

The jumpers can be attached in many buys, as seen in the images that follow (**Figures 8-28, 29, 30, 30**). They must be soldered properly to their connecting points. Anywhere the wire makes contact with the component lead, a fillet must be seen. After cleaning, the connection should be smooth and free of flux pitting.

Select the proper gauge wire first. The wire should have sufficient area in cmils to safely conduct the electrical load. A rough gauge is to select a wire of the same width (or slightly smaller) as the pin/track you are connecting. A 22 AWG solid wire usually meets this requirement. The wire should be stripped sufficiently to allow maximum contact area with the pin and still have enough insulation to prevent shorts. The overlap should be a minimum of two times the pin width.

Measure the needed length of wire and do all of the stripping before soldering; circuit boards are crowded and do not afford much space to solder one connection and then another without some other part getting in the way. The length must be sufficient to run the wire on an X and Y-axis. All direction changes must be at 90° angles. Leave the jumper wire long enough to provide stress relief.

When you have stripped the wire, bend the stripped end so that it hugs the pin and the insulated section lies flat against the board. The stripped leads may be flattened with duckbill pliers if desired. You can use polymide tape to hold the wire in place during soldering. Add flux to the wire/pin junction and using a properly sized tip (with a very small amount of solder on it for a heat bridge), contact the wire/pin at the base, add more solder to the opposite side of the junction (if needed) and wipe the iron up along the pin. Clean the new junction thoroughly and inspect it under a microscope. Jeweler's loops can also be used for the inspection as long as they have X10 magnification. The opposite side of the wire is done the same way.

The wire will have to be bonded to the board if its length is greater than 1" (25.4mm). The wire is *spot* glued, it is not glued along its entire length. Hot melt glue, tape dots, and quick setting glue are acceptable for bonding. The jumper has to be glued within 1/4" (6mm) of both ends, within 1/4" of all bends, and every 1" of straight run.

Figures 8-28, 8-29, 8-30, and **8-31** show acceptable connection methods.

Here is a list of things that should be done:

1) Run the wire on an XY axis.

2) Leave enough slack for stress relief.

3) Leave enough slack so that the jumper can be moved away from any lands it may cross over. This facilitates ease of maintenance later.

4) Bond the wires as described in the previous paragraph.

5) Jumper wires going through a PTH must have their insulation intact through the PTH and must have an additional insulating sleeve.

6) Jumper wires that are connected to a PTH have to be identifiable (i.e. the PTH cannot be filled to the point that the jumper termination disappears.

Here is a list of things that should be avoided:

1) Do not place multiple jumper wires on any pin.

2) Do not run wires over components, i.e. over an IC on which a jumper wire does not terminate.

3) Do not let a jumper wire touch a heat sink.

4) Do not run wires under components (the footprint) unless no other routing is possible.

5) Jumper wires have to be flush to the board and component to allow removal of the board (pulling a card from a motherboard).

6) Do not use bare wire on jumpers longer than 1".

| **Figure 8-28** | **Figure 8-29** |

Figure 8-30

Figure 8-31

Edge Connectors

Most people are familiar with the circuit cards that slip into sockets on computer motherboards. The electrical contacts on these circuit cards are *edge connectors*, sometimes referred to as fingers. The techniques discussed previously, with slight modifications, can be used to repair these edge connectors quite effectively. **Figure 8-32** shows an edge connector with a missing finger.

A few general points before we begin. First, the finger shown is bare copper. If the connector you are working on has been tinned, the repair is straightforward. If it is gold plated, principally to prevent oxidation but also to act as a sort of lubricant, a plating system will be required to plate the new

Figure 8-32

A circuit card edge connector with a missing finger.

connector in gold. Plating is not covered in the text. Second, the variety of edge connectors is large. Look closely at **Figure 8-32** and you can see two styles, one where the lead comes straight out of the connector and second where the lead exits obliquely. Finger frames will usually carry a sufficiently wide assortment to cover most contingencies. If you can't find a finger that works, pilfer one from an out of service board.

The first step in the repair is to remove all of the old finger up to the tinning. The copper is meant

148

to be the connector to the socket, not the tinned section. Once it has been removed, find an identical connector and put it into place using polymide tape. Do not cover the section to be soldered. Recall that polymide tapes are used to prevent ESD damage to any active components, such as CMOS devices which are very likely on the card being repaired. Use cf masking or cellophane tapes can cause sufficient static discharge to destroy any components still on the board.

To solder the connector, begin by making sure the solder point on the original track is tinned. You may wish to add a small amount of solder to the original track. Place a little solder on the iron to form a heat bridge and solder the connector while holding it down with a non-destructive tool. Draw the iron towards the old track to achieve proper wetting. The small amount of extra solder to the right of the heel will not affect the connector's performance other than aesthetically. **Figure 8-34** is a picture of the finger after soldering. You can see the small fillet on the new finger.

The finger now needs to be permanently attached to the edge of the board with an adhesive. An off the shelf epoxy will work well. Remember to mix it properly. When you are ready to use the adhesive, gently bend the edge connector up. You do not want to fold it and place a crease anywhere on its length. As soon as the epoxy is in place, lay the edge connector down and put a Teflon block on top of it. Hold all this in place with a clamp. The clamp in **Figure 8-35**, were it wide enough, could have been used to hold the finger down without the Teflon block.

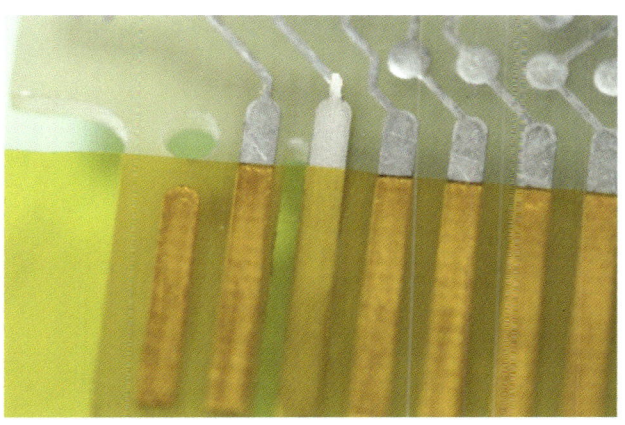

Figure 8-33
Selected edge connector taped down in preparation for soldering.

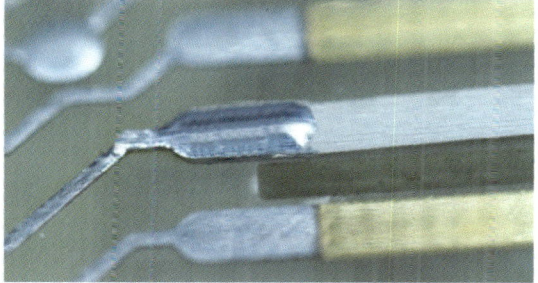

Figure 8-34
A correctly soldered edge connector. The extra solder can be removed but at the risk of removing too much solder or moving the connector.

Figure 8-35
Edge connector clamped in place for drying.

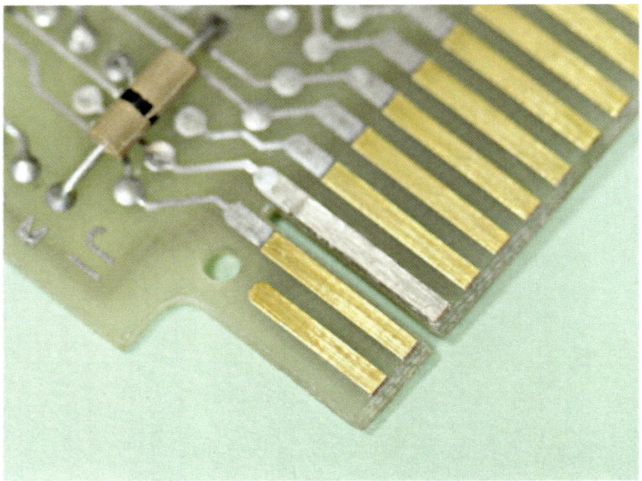

Figure 8-36
A completed edge connector repair.

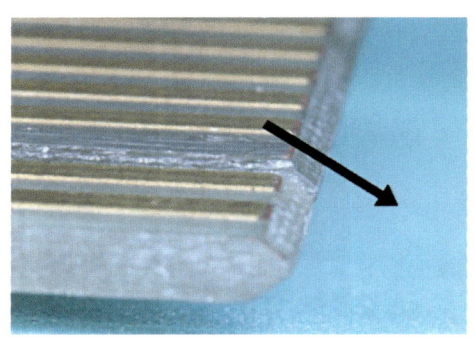

Figure 8-37
Here is the bevel of the edge connector. You want to duplicate that bevel on the new finger and draw the file only in the direction of the arrow.

Once the glue has set, remove everything and inspect. None of the glue should be on the contact. An abrasive stick should be able to remove any excess adhesive. Notice the line you see where the repair overlaps the original track. This should be present if the repair is done properly. The finished product is found in **Figure 8-36.**

Circuit cards with edge connectors usually have a bevel on the insertion end. Use a small precision file to create bevel at the edge of the new finger. File only in the direction indicated by the arrow in **Figure 8-37**. Filing in the opposite direction, towards the board, could cause the pad to lift. If the new finger extends beyond the bevel, filing will remove the excess and align the repair up perfectly with the existing fingers.

It may seem obvious but do not apply a conformal coating to edge connectors. Overzealous technicians have done so in the past with obvious results. Should the board require a conformal coating, use some polymide tape to cover both sides of the edge connection before applying the coating.

Replacing Larger Sections of Damaged PCBs

Three methods of board repair will be discussed in the upcoming sections. The first can be done with the tools already discussed. Unfortunately, the last two are quite delicate and require the use of a mill.

Figure 8-38
Here is a before and after image of a small repair that does not go through the board.

Small Repairs (1/2", 12.5mm or Less): Damage Does Not Extend Through Board

A repair of less than 1/2" (12.5mm) in which the damage does not extend through the board can be made using epoxy. As with the burn repair for a through hole, remove all of the damaged substrate. Once the fiberglass has been removed, make the sides of the repair vertical. Now make an undercut and clean the area thoroughly. Mix the epoxy. Again, minimize the amount of air you mix into the glue. Press the mixed glue between plates as described earlier to remove excess air bubbles. Apply the glue and make sure it gets into the undercut and forms a mound above the surface of the board. Remove any air bubbles with a toothpick. Let the glue set per the manufacturer's data sheet. Should a through-hole be present, be sure to place a Teflon® block behind the hole so glue doesn't run through the opening. Once the adhesive has set, the excess glue can be removed with abrasive bullets and any tracks or pads can be replaced as described in earlier chapters.

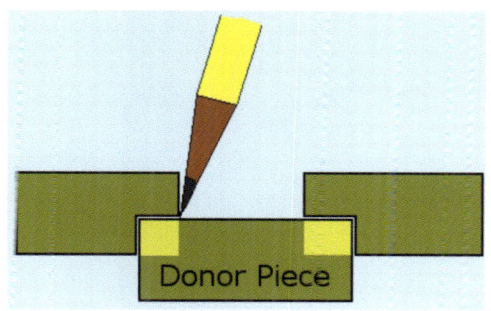

Figure 8-39
A shelf created with a circular saw.

Damage Extends Through Board

Burn repairs that extend through the board can be performed with basic tools. A rotary tool with an assortment of ball mills and a saw will be needed first. Remove all of the damaged material just as you did with a burned pad. We are assuming the damage is extensive, and will leave a hole in the board.

Start by clearing the damage with a ball mill. Try keeping the sides of the excavated area perpendicular to the board's surface. Once the damaged material has been removed, use the saw and cut a shelf in the board (**Figure 8-39**). The shelf does just what the name implies; it forms a surface for the repair material, a section of donor board, to sit on.

The difficult portion of this repair is cutting the donor piece accurately enough to sit on the shelf. Begin by cutting the donor piece from an old circuit card. This can be done with a hand shear, bench shear, keyhole saw, or a scroll saw. Cut the donor substrate slightly larger than the repair area. The donor piece can be filed down to fit the smaller hole in the board requiring repair. Once this is done, flip the board over and, with the donor board under the hole to be repaired, outline the shelf with a pencil (**Figure 8-40**). This gives you an indication of how much material you must remove with the rotary tool and circular saw blade. You may want to attach the rotary tool to a press if one is available. If you fix the saw blade to the depth of the shelf in the hole, you can turn the donor part on the press' base and cut to the

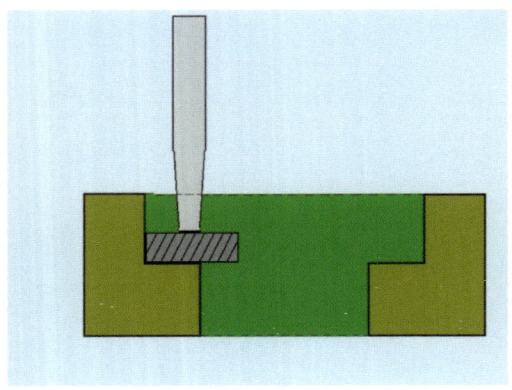

Figure 8-40
Mark the area of the donor board that needs to be removed – the section marked in yellow.

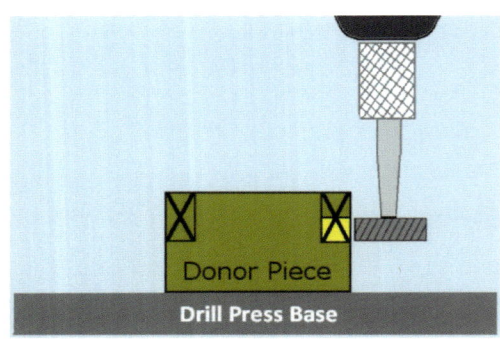

Figure 8-41
Place the piece on a drill press base and fix the drill to a set depth. You can nibble the material away.

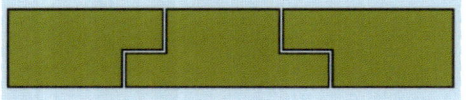

Figure 8-42
With patience, the part can be cut to fit the repair area precisely. Epoxy can be used to finalize the repair.

pencil mark (**Figure 8-41**). Always cut less material than you think you have to. You can always remove more, you cannot put any back. Finally, test fit the piece.

The new piece should fit loosely into the area of the board to be repaired. Epoxy the edges of the donor piece and insert it into the board (**Figure 8-42**). Use a clamp to keep everything secure and parallel. Traces, pads, and through-holes can be added as described in previous sections.

Large Repairs

Substantial board repairs can be done using large sections of substrate donated from an cut of service board or new PCB material (with the copper removed). This method is suitable for replacing damaged edges of a board. However, because of the extreme accuracies required, and the prohibitively expensive equipment needed to achieve these accuracies, the technique will not be covered in depth.

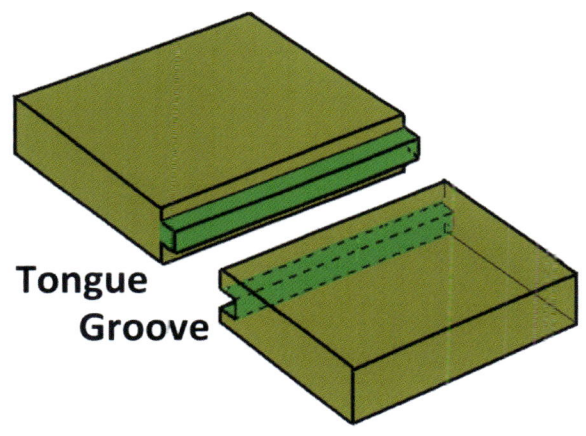

Tongue
Groove

The process begins by removing the damaged edge of the board. A tongue or groove is cut into this piece. A donor board will have a matching tongue or groove cut into it. The two parts undergo a test fit and are trimmed to match. Adhesives are applied to the tongue and groove, and the pieces clamped until cured. All tracks, pads, and through-hole connections can now be installed.

Figure 8-43

Here is a tongue and groove repair of a circuit card. This allows substantial areas to be replaced. The precision required is extreme.

Appendix A
Electrostatic Discharge (ESD)

The control of static electrical discharge has become a priority in the electronics industry - the Electrostatic Discharge Association (ESDA) was formed to deal solely with this issue. Static discharge costs the electronic and computer industries millions of dollars every year in damaged circuits and/or scrambled data. This damage may be seen immediately or it may not become evident for some time. This appendix will deal with some of the causes of ESD and some remedies that should be implemented by any operator in the field.

What is ESD?

ESD is a rapid flow of current between two objects that contact each other or come in close proximity to one another. Current flows in the form of electrons moving between an object with negative potential and one with a positive potential. Everyone is familiar with static electricity – the sudden jolt received when touching a door knob after walking on a carpet. The voltage in this small jolt is at least 3000V. We say "at least" because this is the approximate threshold of sensitivity for most people. In other words, you don't feel the static discharge until a minimum of 3000 is reached. This may appear to be a substantial voltage but values of 35,000V can be achieved by simply walking on a carpet in 10% to 20% humidity environment. It is normal for a person to have a 300V potential difference between their feet and head.

The amount of voltage that can be accumulated depends on the *triboelectric* capacity of an object. The triboelectric charge is the tendency of a material to gain or lose electrons. Look at the table on the following page. Hands have a tendency to be positive in charge, they give up electrons. Teflon prefers to have an excess of electrons. Given these conditions, a difference in potential can occur and lead to a substantial static charge. The further two materials on the triboelectric series are apart; the easier it is for a voltage to accumulate. The drier the environment is, the greater the static build up. High humidity environments decrease static voltage levels to less than $1/10^{th}$ of the low humidity values. One final word on the table – the position of the materials is not absolute. The purity of each substance has a dramatic effect on its position in the triboelectric table. For example, steel and cotton can reverse their positions in the table if they contain a large number of impurities.

Triboelectric Series	
Hands and Skin	Charges most positive
Asbestos	
Glass	
Hair	
Nylon	
Wool	
Mica	
Silk	
Aluminum	
Cotton	Neutral
Steel	
Lucite (Plexiglas)	
Amber	
Balloon	
Mylar	
Copper	
Gold and Platinum	
Rayon	
Polyester	
Styrofoam	
Cellophane Tape	
Polyurethane	
Teflon	
Silicone Rubber	Charges most negative

Table 2-1

ESD and Electronics

Put simply, ESD and electronic components do not mix well. Electrostatic Discharges can have several detrimental effects on electronic and microelectronic devices. For example:

1) ESD can cause a condition called *latch-up*. The sudden surge voltage causes a portion of a circuit to stay in one state, a high or a low. The circuit effectively becomes inoperative. This latch-up condition can occur as soon as the ESD event happens or the failure might not occur for some time.

2) The junction on a component, a **C**omplementary **M**etal **O**xide **S**emiconductor (CMOS) chip for example, can have the ESD spike burn a hole through the insulating gate material. The component then leaks or becomes inoperative.

3) ESD can cause the metal in a semiconductor, such as the CMOS chip just mentioned, to melt and migrate through the insulator. Once again, the effect is a leaky or inoperative component.

Blowing a hole through an insulator or melting metal may seem to be an impressive feat. In some ways, it is but the miniaturization which makes electronics so functional also makes such ESD damage possible. The stylized drawing in **Figure 2-1** is of NMOS device (related to the CMOS family). Its size on a chip could be as small as 15 to 20 microns (15 to 20 *millionths* of a meter). The oxide layer must be substantially smaller than this, perhaps on the order of just a few atoms. It would be not be difficult for a few tens of volts to penetrate such a thin layer. Some electronic devices can be damaged by as little as 10 volts. If you recall, the *normal* potential difference on a person is about 300V. Just touching this device could destroy it.

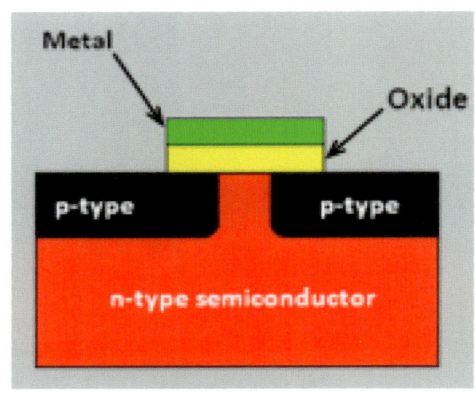

Figure 2-1

The Things We Do

The things we commonly do in electronics can lead to conditions conducive to static discharges. Walking across a floor has already been mentioned. Here are a few more possible troublemakers.

1) Using clear cellophane tapes

2) Transporting electronics in ungrounded carts

3) Using high pressure air to clean components

4) Storing or shipping parts in Styrofoam containers

5) Sitting in a chair with urethane foam

6) Co-workers

7) Hand tools (de-soldering tools)

8) Soldering irons

9) Walking across a waxed floor

10) Wearing nylon clothing

11) Taking out the trash – pulling the plastic trash bag out of a plastic bin.

Some of the items listed are obvious, some may be surprising. The point to be made is that as a maintainer, know what you are working on (how ESD sensitive is it) and know how to eliminate ESD causing elements.

Controlling ESD

Technicians have an immediate impact on ESD control. *As a minimum, wear an ESD wrist strap and perform all work on anti-static mats.* The mats and wrist straps make contact with the skin and work item and provide a path to ground for stray electric charges. The straps and mats do not directly ground the user – this would be quite unsafe. Instead, there is a resistance of 1,000,000 ohms (1MΩ) between the technician and earth ground. This resistance is present to limit the current to safe levels. If the technician had the usual 300V potential on him, the 1MΩ resistance would limit the current to 0.0003A (300µA) – a safe level. The grounding mats do exactly the same thing; they slowly drain off the charge at a safe current level.

The work area must also provide ESD remedies. There must of course be a good grounding point for the ESD wrist straps, foot straps, and work mats. This grounding point should be tested by the facility owner to ensure it is safe and effective. An area with a test point to check wrist and foot straps should also be available – these items can be damaged or lose their effectiveness over time.

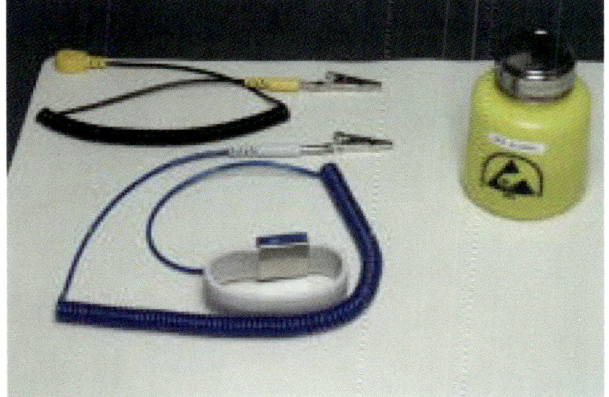

Figure 2-2

Shown here are an ESD mat, ESD wrist strap and an ESD approved solvent dispenser.

Other measures that can be taken are:

1) Monitor and control humidity. It is best to maintain the levels between 40% and 60 %. The 40% level reduces the static build up while the 60% limit is usually the upper range most people find comfortable.

2) Use ESD packaging to contain and transport items.

3) Make sure the floor has an ESD finish – it should provide 100KΩ to 100GΩ resistance depending on application.

4) Make sure any gloves or finger cots are ESD approved.

5) Any tapes used must be a polyimide tape.

6) All test equipment must be ESD safe.

7) Soldering irons, solder extractors, brushes, swabs, etc must be ESD approved.

8) Where an item cannot be properly grounded, use air ionizers. These work by adding a charge to the air particles. These charged particles seek out their opposite charges on both conductors and insulators and neutralize them. Ionizers can be used to make mini-clean room areas (a single work-station) or be large enough to control static in large facilities.

10) Solder fume extraction equipment should be ESD approved if necessary.

11) Keep all people that are not ESD safe at least four feet away from the work area.

12) Use ESD smocks to reduce body charge.

13) ESD lotions can also be used for added security.

14) This is the big one – *TRAINING*. It is impossible to implement an ESD program if the people to whom it applies do not understand ESD and its consequences. Know the problem and the solution.

CPSIA information can be obtained at www.ICGtesting.com
Printed in the USA
LVIW01n1652060915
453034LV00005B/45